New Approach to Genetic Diseases

Academic Press Rapid Manuscript Reproduction

New Approach to Genetic Diseases

Edited by

Takehiko Sasazuki
Medical Institute of Bioregulation
Kyushu University
Fukuoka, Japan

1987

ACADEMIC PRESS
Harcourt Brace Jovanovich, Publishers
Tokyo Orlando San Diego New York
Austin Boston London Sydney Toronto

ACADEMIC PRESS JAPAN
Ichibancho Central Bldg., 22-1 Ichibancho, Chiyoda-ku, Tokyo 102

United States Edition published by
ACADEMIC PRESS, INC.
Orlando, Florida 32887

United Kingdom Edition published by
ACADEMIC PRESS, INC. (LONDON) LTD.
24/28 Oval Road, London NWI 7DX

Library of Congress Cataloging-in-Publication Data

New Approach to Genetic Diseases

Proceedings of the International Symposium, "New Approach to Genetic Diseases" sponsored by the Japan Intractable Diseases Research Foundation, held in the summer of 1986 in Tokyo, Japan.

Includes index.
I. Genetic disorders--Congresses. I. Sasazuki, Takehiko.
II. International Symposium "New Approach to Genetic Diseases" (1986 : Tokyo, Japan) [DNLM : 1. DNA--diagnostic use--congresses. 2. Genetics, Medical--congresses.
3. Hereditary Diseases--congresses. 4. Immunogenetics--congresses.
QZ 50 N5317 1986]
RB155. N487 1987 616'. 042 87-72457
ISBN 0-12-619415-7 (alk. paper)

Printed in Japan
87 88 89 90 9 8 7 6 5 4 3 2 1

Contents

IV. DNA DIAGNOSIS AND GENE THERAPY

Preface

Since the 1970s, the armamentarium available to molecular immunologists and geneticists has increased considerably and new insights have been gained concerning the mechanisms involved in the pathogenesis of monogenic and multifactorial disorders. Molecular cloning and determination of the base sequences of genes, monoclonal antibody techniques, transformation of single cells and related events *in vivo* by inserting foreign genes have shed light on the pathogenesis and pathophysiology of disease. The biometrical analysis of family and population date made similar advances, albeit with a lesser degree of fanfare. In several multifactorial disorders, major and/or submajor genes with close linkage to a disease have been identified. Multifactorial diseases were heretofore thought to be under the control of a "polygene" which could not be given attention at the molecular level. Once the presence of a major or submajor gene responsible for the multifactorial disease is made known, the new technology can be applied for elucidation.

The International Symposium, "New Approach to Genetic Diseases" was held in Tokyo in the summer of 1986. This meeting, supported by the Japan Intractable Diseases Research Foundation had participants from 11 countries. These geneticists, immunologists, endocrinologists, embryologists, molecular biologists and clinical scientists discussed molecular mechanisms involved in the pathogenesis of intractable diseases, including rare Mendelian monogenic disorders and common multifactorial diseases. DNA, used as a diagnostic, and also gene therapy were highlighted topics of discussion.

An earthquake greeted the participants in Tokyo, many of whom had never experienced earthquake tremors. Perhaps this "natural" event will serve as a remembrance of the Tokyo meeting, the purpose of which was to examine the present stage in the search for causes of intractable disease.

We hope this volume will provide stimulation for thought——and action.

Takehiko SASAZUKI, M. D.
Summer 1987

I . GENETIC ANALYSIS OF COMMON DISEASES

A GENETIC ANALYSIS OF PREMATURE CORONARY ARTERY DISEASE

David J Galton
Medical Professorial Unit
St Bartholomew's Hospital
London, UK

INTRODUCTION

Coronary atherosclerosis - one of the major health problems of our time - arises from a complex interaction between environmental and genetic factors. Evidence for genetic factors comes from many studies reporting a strong familial aggregation of the disease. A heritability (that is the proportion of phenotypic variance that can be attributed to additive genetic factors) of 0.63 for coronary artery disease of early onset in family studies has been reported (1). Twin studies have revealed a higher concordance rate of the disease in monozygotic than in dizygotic pairs. In one representative study of coronary artery disease occurring before the age of 60, concordance rates were 0.83 and 0.22 for mono- and dizogotic twin pairs respectively, the disease end point being angina pectoris and/or myocardial infarction (2).

There is now a mass of clinical, epidemiological and experimental data to implicate altered metabolism of lipoproteins in the aetiology of atherosclerosis. In rare instances a simple monogenic disorder such as a defect in the cell-surface receptor for endocytosis of LDL can strongly predispose to the development of premature atherosclerosis (3). However in the majority of cases it is not possible to identify single genetic determinants and it is likely that several major genes may contribute to the manifestation of the disease. It will be of great interest to map which of the major genes segregate with the trait. This may allow identification from birth of individuals who are predisposed to develop premature atherosclerosis.

The most likely candidates for genetic variants predisposing to atherosclerosis are those coding for the lipid transport proteins; and the next section summarises recent information on the apolipoprotein gene family.

The Apolipoprotein Genes

Use of recombinant DNA techniques has identified the chromosomal locations of all the major apolipoprotein genes and permited a detailed study of their genomic structure. The apolipoprotein genes are dispersed amongst 4 chromosomes. There is a cluster of three on the long arm of chromosome 11 (apo A1, CIII and AIV). Another cluster occurs on chromosome 19 (apo C1, CII and E; together with the LDL receptor). Two others are located on chromosome 1 (apo AII) and on the short arm of chromosome 2 (apo B).
There are considerable nucleotide sequence homologies amongst the apoproprotein genes, and it is possible they arose from a common ancestral gene by duplications or conversion events. Thus their genomic structures reveal the presence of three introns in strikingly similar positions. One is within the 5' untranslated region of the gene; the second within the leader sequence; and the third intron within the coding part of the gene. These divide the gene into four exons; and the fourth exon is variably expanded by internal repeat units of approx. 66 base pairs. These may code for aminoacid sequences that form the amphipathic helix and be responsible for the lipid binding properties of these apolipoproteins. A possible scheme for the evolution of this multigene family is presented in (5). The common ancestral gene may have been very similar to the present day apolipoprotein C-1 in structure and length. Then by a series of gene duplications the other genes were formed and diversified by variable duplication of stretches of codons within their fourth exons.

Genetic Polymorphisms

Mutations are a fundamental way of altering the genome and in some instances can produce a deleterious change in gene function. Such mutations may gradually become eliminated from the gene pool of the population under the influence of natural selection on the resulting phenotype. However occasionally a mutation will arise that may have some selective advantage over the common allele and may gradually increase in frequency to the point of permanent fixation in the population (4). When the frequency of the uncommon variant occurs to an appreciable extent that cannot be accounted for by new mutations (arbitrarily taken as more than 1%) it is called a polymorphic variant. A genetic polymorphism is therefore the occurrence in the same population of two or more alleles at one locus, each occurring at an appreciable frequency and not accounted for by new mutations. The DNA polymorphism may not occur within structural genes. It has been calculated (6) that one in every hundred nucleotides may be variable in flanking regions of the gene or in spacer regions (ie intergenic DNA). This makes the extent of DNA polymorphisms much greater than observed for protein polymorphisms. Genetic polymorphisms constitute a basic component of hereditary variation in natural populations on which evolutionary forces operate. A polymorphic variant that is of minor selective disadvantage, and may be declining in frequency may become a selective advantage if the prevailing environmental conditions alter slightly. This therefore provides a means whereby populations can evolve in different directions. In general the greater the genetic variation inherent in a natural population the greater the opportunities for evolutionary change.

From a preliminary analysis (human genomic probes have only been available since the mid-1970s) there is clearly a large amount of genetic polymorphism in human populations. Some of this polymorphism may be impermanent representing a transient phase of molecular evolution. The rarer of the variant alleles could be in the process of increasing or decreasing in frequency, under the influence of natural selection and/or by random genetic drift. Alternatively the polymorphism may be permanent ie in equilibrium with the common form.

Detection of Polymorphisms

DNA polymorphisms can be detected if the mutation directly affects the recognition site of a restriction endonuclease, either creating a new site or abolishing an old one. In either case a different sized DNA fragment will be produced. More than 100 restriction enzymes are available each with different nucleotide sequence specifities and each enzyme is of potential use for detecting new polymorphisms. This is currently providing a wealth of new genetic markers for clinical studies. Alternatively a DNA polymorphism can be produced by variable insertions of repetitive sequences within a short stretch of the genome. This leads to DNA fragments produced by digestion with restriction enzymes of highly variable length.

The most useful method to detect DNA polymorphisms is by "Southern blotting" (17). Usually leucocyte DNA is digested with various restriction enzymes and fractionated according to size on agarose or polyacrylamide gel. The DNA on the gel is denatured with alkali, neutralized with Tris buffer and transferred by capillary flow to a nitrocellulose or other filter - the blotting process. The filter now carries a "print" of the DNA from the gel that can be hybridised to a labelled gene probe and then visualised by autoradiography. This probe can be a cloned DNA sequence or a synthetic oligonucleotide. Using different restriction enzymes this method can generate far more allelic variants than is possible by a study of protein polymorphisms. It is now possible to take a gene which has no protein variants associated with it and in a relatively short time discover several DNA variants at that locus. Such DNA variants may be due to mutations in introns, exons, flanking sequences or spacer DNA. The latter can be in linkage disequilibrium with nucleotide variation within the coding region of structural genes and can be of diagnostic value.

Genetic Polymorphisms related to Atherosclerosis

An early locus to be studied for population assocations with coronary artery disease was the apo A1-CIII-AIV gene cluster on the long arm of chromosome 11. This region appears to be relatively prone to new mutations and DNA rearrangements. The apo A1 and AIV genes (that are closely related) have been separated by the interposition of the apo CIII gene in the opposite orientation to both and intra-strand recombination has been described between the A1 and CIII genes. There are also multiple restriction site polymorphisms along the genome over a distance of approximately 14Kb (7). These occur in exons, introns and flanking sequences as described in Fig 1. These sites reveal a complex pattern of linkage associations. Linkage analysis has been done using haplotypes of pairs of allelic sites. The degree of linkage disequilibrium is not a simple function of the relative physical distance between sites over a region of 14 Kb. For example the Sst1 and Pvu II restriction site polymorphisms are close together yet in linkage equilibrium; the Sst 1 and Msp 1 sites are further apart yet show linkage equilibrium. The simplest explanation is that the processes of mutation and genetic drift are primarily responsible for the observed associations. The relative role of differential selection on haplotypes is difficult to assess. The effects of recombination over such short distances are potentially less significant than mutation and drift/selection in the formation of new haplotypes (9).

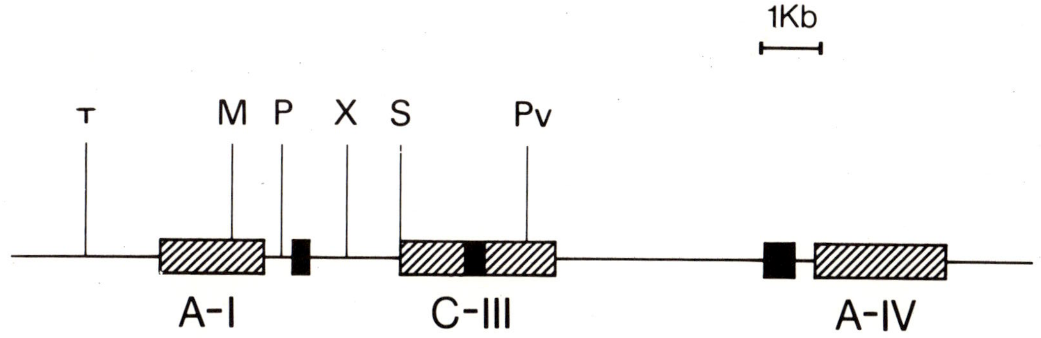

Fig 1: Restriction map of the A1/CIII/AIV genes showing multiple polymorphic sites. Hatched bars are exons/introns and spacer DNA are lines. Solid boxes are repetitive elements. Abbreviations for restriction endonucleases: Pv = Pvu II; S = Sst 1; X = Xmn 1; P = Pst 1; M = Msp 1; T = Taq 1.

Apo C-III Polymorphisms

Exon 4: A DNA polymorphism for the restriction enzyme Sst 1 occurs in the fourth exon 39 base pairs from the C-terminal codon (10). It is due to a C to G transversion and can be used to identify two alleles, S1 and S2 by Southern blotting. The frequency distribution of this polymorphism has been extensively studied in different ethnic populations as well as in disease groups (11). In a study of coronary artery disease defined by coronary angiography (12) the frequency of the rare allele (S2) rose from 6% (n=68) in the controls to 22% (n=61) in severely affected subjects (Fig 2). A similar study has been performed with young survivors of myocardial infarction (13). The variant (S2) allele was found in 4% (n=4) of healthy controls compared to 21% (n=41) of infarction subjects (fig 2).

Intron 1: There is a Pvu II site polymorphism in the first intron of the apo CIII gene giving rise to two alleles of respective frequencies 0.79 and 0.21 in a West German population (8). The frequency of the rare allele is increased in subjects (n=155) with coronary artery disease as defined by angiography compared to healthy controls with a relative incidence of 3.5 (calculated by the method of Woolf).

The Apo A1 Gene

This gene is situated 2.6 Kb upstream from the apo CIII gene and is transcribed in the opposite orientation. There is a Pst 1 site polymorphism 314 base-pairs from the 3' end of the apo A1 gene. The frequency of the rare (P2) allele in randomly selected controls was 4% (n=123) and was 3.3% in 30 subjects with no evidence of coronary disease by angiography. By contrast in 88 patients with coronary disease prior to the age of 60 years (defined by angiography) the P2 allele occurred in 31.8% ($p<0.0001$) of subjects (14). This at present is the strongest genetic marker reported for coronary artery disease.

The next stage in these studies is to examine haplotype combinations of polymorphic restriction sites to see if they constitute stronger genetic markers for atherosclerosis (15). The A1-CIII-AIV cluster with multiple restriction site polymorphisms can provide valuable genetic markers since polymorphic sites that are physically close together can show a number of haplotype combinations (18). These can act as multi-allelic markers and may provide better identification of any putative atherogenic loci in their vicinity.

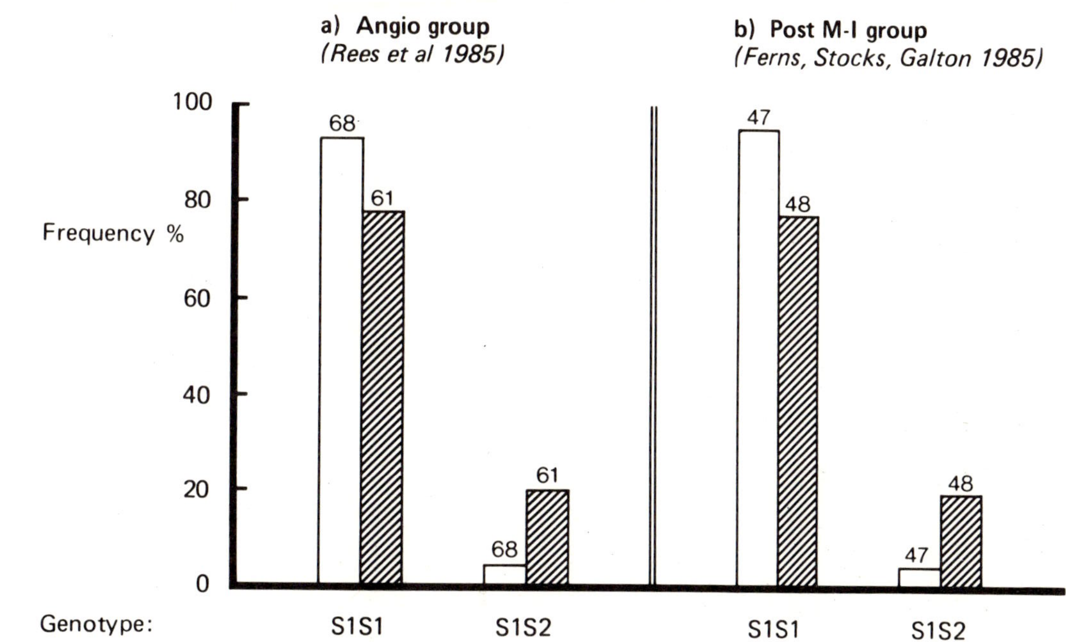

Fig 2: Distribution of an Sst 1 polymorphism in the apolipoprotein CIII gene in early-onset coronary artery disease. Genotype frequencies are presented in patient groups with either severe or minimal coronary artery disease defined by angiography (see reference 12), or in survivors of myocardial infarction (see reference 13). Open bars, minimal disease; hatched bars, severe disease.

A possible interpretation of the above studies is that these DNA polymorphisms at the AI-CIII locus are harmless mutations but are acting as linkage markers for a deleterious atherogenic allele in this region. The reasons why the linkage markers vary amongst three populations (West Germany, Britain and North America) could be that:- (1) the atherogenic allele has mutated several times during evolution and hence associates with different background polymorphisms in different populations as a result or (2) there have been recombinational events between the atherogenic allele and the different polymorphic restriction sites. The mutation therefore can spread to a different background of chromosomal polymorphisms. The latter is somewhat unlikely in view of the short physical distances between the various linkage markers.

There are also other examples in the apoprotein gene family where multiple mutations have probably occurred. For example the apoprotein CII gene mutation producing familial apo C-II deficiency is linked to one allele of a Taq 1 polymorphism in a North Italian pedigree, but to the other allele in a Dutch pedigree (16). The apo C-II gene has probably mutated twice to account for these different allelic associations. A similar situation may have arisen with the putative atherogenic allele on the long arm of chromosome 11.

Conclusion

These studies represent first attempts to identify major alleles that segregate with the inheritance of premature coronary artery disease. Identification of these atherogenes may allow prediction from an early age of individuals who are predisposed to develop premature atherosclerosis. Thereafter treatment could be instituted to modify environmental factors that interact with such high risk genes to produce disease. One of the most important applications of this area of genetic research is a more precise appreciation of the environmental factors that interact with the genome to predispose to atherosclerosis. Finally information on the major atherogenic loci may lead to newer forms of therapy directed against possible deleterious gene products.

ACKNOWLEDGEMENTS

Financial support for these studies have come from the Medical Research Council, Welcome Trust and British Diabetic Assocation which is gratefully acknowledged. Financial support from the Fritz-Thysson Foundation is also gratefully acknowledged.

REFERENCES

1. Nora JJ, Lortscher RH, Spangler RD, Nora AH, Kimberley WJ (1980) Genetic-epidemiologic study of early-onset ischaemic heart disease. Circulation 61:503-508.

2. Berg K (1983) Genetics of coronary heart disease in "Progress in Medical Genetics vol V" eds. Steinberg, Bearn, Motulsky and Childs. publ WB Saunders & Co.

3. Brown MS, Goldstein JL (1976) Receptor mediated control of cholesterol metabolism. Science 191:150-154.

4. Galton DJ (1985) Molecular genetics of common metabolic disease. publ. E Arnold.

5. Luo CC, Li WH, Moore MN, Chan L (1986) Structure and evolution of the apolipoprotein multigene family . Journal Molecular Biology 187:325-340.

6. Jeffreys AJ (1979) DNA sequence variants in the G-, A-, S-and B-globin genes of man. Cell 18:1-10.

7. Breslow J L (1985) Human apolipopoteins : molecular biology and genetic variants. Annual Review Biochemistry 54:699-727.

8. Frossard PM, Coleman R, Funke H, Assmann G (1986) Molecular genetics of the human apo Al-CIII-AIV gene complex : application to detection of susceptibility to atherosclerosis (in press).

9. Barker D, Holm T, White R (1984) A locus on chromosome 11 p with multiple restriction site polymorphisms. Am J Human Genetics 36:1159-1171.

10. Rees A, Shoulders CC, Stocks J, Galton DJ, Baralle FE (1983) DNA polymorphism adjacent to the human apoprotein A-1 gene Lancet ii;444-447.

11. Rees A, Sharpe C, Stocks J, Shoulders CC, Baralle FE, Galton DJ (1985) DNA polymorphism in the apo Al/CIII gene cluster. Journal Clinical Investigation 76:1090-1095.

12. Rees A, Stocks J, Williams LG, Caplin JL, Jowett NI, Camm AJ, Galton DJ (1985) DNA polymorphisms in the apolipoprotein CIII and insulin genes and atherosclerosis. Atherosclerosis 58: 269-275.

13. Ferns GAA, Stocks J, Ritchie C, Galton DJ (1985) Genetic polymorphisms of apolipoprotein CIII and insulin in survivors of myocardial infarction. Lancet i:300-303.

14. Ordovas JM, Schaefer EJ, Salem D et al (1986) Apolipoprotein Al gene polymorphism associated with premature coronary artery disease and familial hypoalphalipoproteinaemia. New England J Medicine 314:671-677.

15. Rees A, Stocks J, Paul H, Ohuchi Y, Galton DJ (1986) Haplotypes identified by DNA polymorphisms at the apolipoprotein Al/CIII gene locus and hypertriglyceridaemia. Human Genetics 72:168-171.

16. Humphries SE, Williams LG, Stalenhoef AF, Baggio G, Crepaldi G, Galton DJ, Williamson R (1984) Familial apolipoprotein CII deficiency : a preliminary analysis of the gene defect in three independent individuals. Human Genetics 67:151-156.

17. Southern E M (1975) Detection of specific sequences among DNA fragments separated by gel electrophoresis. J Molecular Biology 98:503-514.

APOLIPOPROTEIN AI-CIII GENE POLYMORPHISMS AND MYOCARDIAL INFARCTION IN A JAPANESE POPULATION

Hideo Hamaguchi[1], Naoko Hattori[1], Junichi Satoh[1], Mieko Onuki[1], Kimiko Yamakawa[1], Hideomi Fujiwara[3], Hiroshi Amamiya[3], Hideo Nagaoka[3], Yasuko Yamanouchi[1], Takaaki Okafuji[1], Yukio Iwamura[1], and Shigeru Tsuchiya[2]

[1]Institute of Basic Medical Sciences and
[2]Institute of Community Medicine,
University of Tsukuba,
Ibaraki-ken, Japan
[3]Tsuchiura-kyodo Hospital,
Tsuchiura, Japan

INTRODUCTION

Apolipoprotein (apo) AI is the major protein constituent of high-density lipoprotein (HDL) (1). Apo CIII is a component of chylomicrons, very-low-density lipoproteins (VLDL) and HDL (1). The human apo AI and apo CIII genes are tightly linked and form a gene complex together with the apo AIV gene on the long arm of the chromosome 11 (2-6). A mutation of the apo AI gene is associated with low plasma HDL and premature atherosclerosis (7, 8). In addition, restriction fragment length polymorphisms related to the apo AI and apo CIII genes have been reported to be associated with coronary atherosclerosis and familial hypoalphalipoproteinemia in Caucasians (9-14). An apo AI-CIII gene polymorphism associated with coronary atherosclerôsis may be a linkage marker for a deleterious atherogenic allele at a locus in

Supported by a Scientific Research Grant from the Ministry of Education, Science and Culture, No.61571088, and a Research Grant for Intractable Diseases from the Ministry of Health and Welfare, Japan the apo AI-CIII-AIV gene complex. Such a linkage marker is likely to vary with the race.

This study has been carried out to search for a linkage marker for the putative atherogenic gene in the apo AI-CIII-AIV gene complex in Japanese. We have determined apo CIII *Sst*-I genotypes and apo AI *Msp*-I genotypes in Japanese myocardial infarction survivors and healthy subjects, because Rees *et al.* have recently reported that the haplotype *S1-M2* defined by *Sst*-I and *Msp*-I polymorphisms is associated with hypertriglyceridemia in Japanese (15). Our data suggest that the haplotype *S1-M2* is also associated with coronary atherosclerosis in Japanese.

MATERIALS AND METHODS

Blood was collected from 69 Japanese myocardial infarction survivors without diabetes mellitus (59 males, 10 females) attending the cardiac clinics at Tsuchiura-kyodo Hospital and Tsukuba Medical Center, Ibaraki-ken. Diagnosis was confirmed by clinical history, typical electrocardiographic changes and enzyme examination. The mean (±SD) age of the myocardial infarction survivors was 58±10. The control group consisted of 82 healthy members of the university staffs and students.

DNA was isolated from whole blood cells essentially according to the method of Kunkel *et al.* (16). Ten μg DNA were digested with 60 units *Sst*-I or *Msp*-I using assay conditions specified by the manufacturers (*Sst*-I, BRL; *Msp*-I, Nippon Gene Co., Ltd., Toyama). Digests of DNA were electrophoresed on 0.85% or 1.0% agarose gels and transferred onto nitrocellulose filters by Southern blotting (17). The filters were hybridized with a ^{32}P-labelled human apo A clone comprising 2.2 kb *Pst*-I fragment of an apo AI genomic clone (18). Gene fragments hybridizing to the probe were detected by autoradiography and fragment sizes were determined by running *Hin*d-III digested lambda phage fragments with each batch of digests (19, 20).

RESULTS

The Polymorphisms

Figure 1 shows representative autoradiograms together with a simplified restriction site map of the apo AI-CIII genes. The *S1* and *S2* alleles were characterized by the frag-

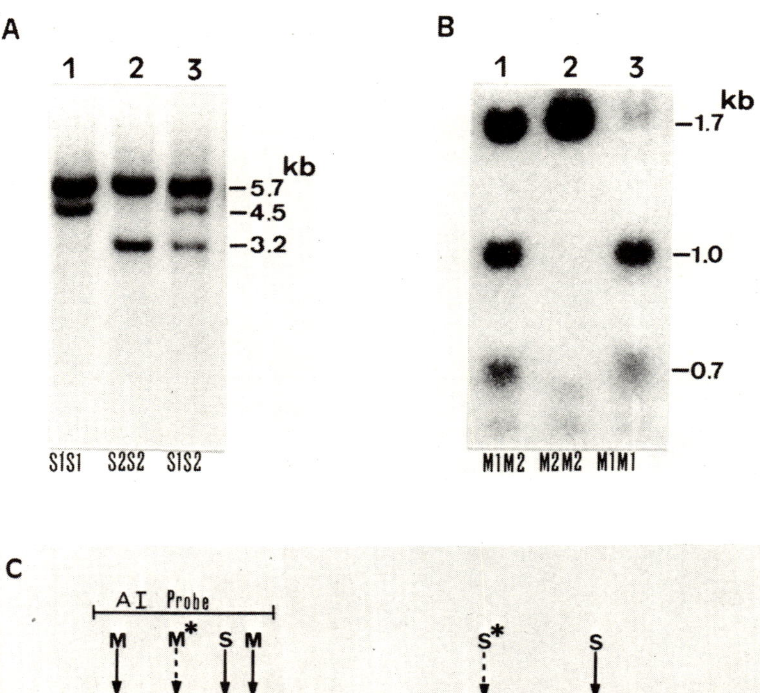

Figure 1. Autoradiograms of Apo CIII *Sst*-I Genotypes (A) and Apo AI *Msp*-I Genotypes (B), and a Simplified Restriction Site Map of the Apo AI-CIII Genes (C). S* and M* represent the polymorphic sites for the enzymes *Sst*-I and *Msp*-I, respectively.

ments of 4.5 kb and 3.2 kb, respectively. The *M1* allele was characterized by the 1.0 and 0.7 kb bands and the *M2* allele by the 1.7 kb band.

Tables 1 and 2 present the distribution and allele frequencies of the apo CIII *Sst*-I genotypes and apo AI *Msp*-I genotypes, respectively, in myocardial infarction survivors and controls. Both the frequencies of the *S2* and *M2* alleles in controls are similar to those for 35 normolipidemic Japanese reported by Rees *et al.* and much higher than in Caucasians (11, 15). Also in controls, the observed numbers are close to those calculated on the assumption that the Hardy-Weinberg equilibrium is observed (for the apo CIII *Sst*-I genotypes, $\chi^2=0.50$, df=2, $0.70<P<0.75$; for the apo AI

Table 1. Distribution and Allele Frequencies of the Apo CIII *Sst*-I Genotypes in Myocardial Infarction (MI) Survivors and Controls

	N	Genotype			Allele	
		S1S1	*S1S2*	*S2S2*	*S1*	*S2*
MI Survivors						
Observed	69	28	38	3	0.68	0.32
Expected	69	32.0	30.0	7.0		
Controls						
Observed	82	37	34	11	0.66	0.34
Expected	82	35.7	36.8	9.4		

Expected values calculated from allele frequencies.

Table 2. Distribution and Allele Frequencies of the Apo AI *Msp*-I Genotypes in Myocardial Infarction (MI) Survivors and Controls

	N	Genotype			Allele	
		M1M1	*M1M2*	*M2M2*	*M1*	*M2*
MI Survivors						
Observed	69	20	37	12	0.56	0.44
Expected	69	21.5	34.0	13.5		
Controls						
Observed	82	32	35	15	0.60	0.40
Expected	82	29.5	39.5	13.1		

Expected values calculated from allele frequencies.

Msp-I genotypes, χ^2=0.96, df=2, 0.50<*P*<0.70). As to survivors of myocardial infarction, the observed and expected values are similar to each other for the apo AI *Msp*-I genotypes (χ^2=0.53, df=2, 0.70<*P*<0.75), but there seems to be heterogenity between observed and expected values for the apo CIII *Sst*-I genotypes though the difference is not significant (χ^2=4.92, df=2, 0.05<*P*<0.10). Unlike the association of the *S2* and *M2* alleles with myocardial infarction found in Caucasians (9-11), there were no differences in both the frequencies of the *S2* and *M2* alleles between Japanese myocardial infarction survivors and controls.

The Haplotypes

The presence of linkage disequilibrium between the *S2* and *M2* alleles has been reported in both the Caucasians and Japanese: all the *S2* allele was observed to be associated with the *M2* allele (11, 15). Indeed none of the 153 subjects examined in this study had the following genotypes: *S1S2M1M1*, *S2S2M1M1*, and *S2S2M1M2* (Tables 3 and 4). As given in table 3, the distribution of the apo CIII *Sst*-I and A-I *Msp*-I genotypes in controls is compatible with there being three of the four possible haplotypes: the haplotype frequencies are

Table 3. Observed and Expected Genotype Frequencies Assuming Linkage Disequilibrium or Equilibrium in Controls

Genotype	Observed (N)		Expected			
			Disquilibrium[a]		Equilibrium[b]	
S1S1M1M1	32		29.9		14.4	
S1S2M1M2	30		33.8		14.5	
S2S2M2M2	11		9.6		2.0	
S1S1M1M2	5		5.4		15.8	
S1S2M2M2	4	4	3.1	3.3	6.2	36.3
S1S1M2M2	0		0.2		6.8	
S1S2M1M1	0		0		13.3	
S2S2M1M1	0		0		4.3	
S2S2M1M2	0		0		4.7	
Total	82		82		82	

[a]Expected values calculated from haplotype frequencies: *S1-M1*=0.604, *S2-M2*=0.341, *S1-M2*=0.055.
[b]Expected values calculated from gene frequencies: *S1*=0.66, *S2*=0.34; *M1*=0.60, *M2*=0.40.

S1-M1=0.604, *S2-M2*=0.341, *S1-M2*=0.055 ($\chi 2$=1.21, df=4, 0.80<P<0.90). The delta value (Δ) calculated using the haplotype frequencies are 0.206±0.059 (P<0.001). In contrast, the distribution of the expected values assuming equilibrium between the two polymorphic sites was highly significantly different from the observed (χ^2=98.7, df=4 or χ^2=120.7, df=8, after Yates' correction; P<0.001).

Table 4 shows the distribution of the combined genotypes in myocardial infarction survivors and controls. The frequency of the individuals with the haplotype *S1-M2* (that is genotypes *S1S1M1M2*, *S1S2M2M2*) in controls is 11.0% and very similar to that (11.1%) for 35 normolipidemic Japanese reported by Rees *et al.* (15). The individual with haplotype *S1-M2* is significantly increased in myocardial infarction survivors (24.6% versus 11.0%, P<0.05).

Table 4. Distribution of Combined Genotypes in Myocardial Infarction (MI) Survivors and Controls

Combined Genotype	MI Survivors		Controls	
	N	(%)	N	(%)
S1M1 / S1M1	20	52(75.4)*	32	73(89.0)
S1M1 / S2M2	29		30	
S2M2 / S2M2	3		11	
S1M1 / S1M2	8	17(24.6)*	5	9(11.0)
S2M2 / S1M2	9		4	
Total	69	(100)	82	(100)

*χ^2=4.90, df=1, P<0.05, when compared with controls.

DISCUSSION

It has been reported that both the *S2* and *M2* alleles are relatively uncommon and associated with coronary atherosclerosis in Causasians (9-11). In the case of Japanese, however, the result of the present study indicates that both the alleles are common and not associated with coronary atherosclerosis. The apo CIII *Sst*-I polymorphism arises from a C-G transversion in the 3' non-coding region of the apo CIII gene (2). The apo AI *Msp*-I polymorphism arises from the presence or absence of a *Msp*-I site in the third intron of the apo AI gene (21). Consequently it is very likely that the *S2* and *M2* alleles may represent linkage markers for atherogenic gene(s) at least in local Caucasians but not in Japanese.

In the present study, it was clearly demonstrated that the alleles identified by the apo CIII *Sst*-I and apo A-I *Msp*-I polymorphisms are in linkage disequilibrium in Japanese. Three of the four possible haplotypes were identified: the haplotype frequencies were *S1-M1*=0.604, *S2-M2*=0.341 and *S1-M2*=0.055 in healthy subjects (Δ=0.206±0.059, P<0.001). On the other hand, the frequency of the haplotype *S1-M2* in myocardial infarction survivors was 0.123 and significantly increased compared with that in healthy adults. The result seems to be consistent with the finding reported by Rees *et al.* (15) that the haplotype *S1-M2* is associated with hypertriglyceridemia in Japanese. As to the *S2* allele in Caucasians, it has been reported to be associated with both of coronary atherosclerosis and hypertriglyceridemia (9-11, 22). Our data suggest that the haplotype *S1-M2* may be a linkage marker for the putative deleterious atherogenic gene in the apo AI-CIII-AIV gene complex in Japanese. However, since the difference in the frequency of the haplotype *S1-M2* between myocardial infarction survivors and healthy subjects was a borderline level, it is required to increase the sample sizes in order to confirm the result of the present study.

SUMMARY

To search for a linkage marker for the putative deleterious atherogenic gene in the apo AI-CIII-AIV gene complex in Japanese, apo CIII *Sst*-I genotypes and apo Al *Msp*-I genotypes were investigated in 69 Japanese myocardial infarction survivors and 82 healthy subjects, using genomic hybridization analysis. The frequencies of the *S2* and *M2* alleles were 0.34 and 0.40, respectively, in healthy subjects. There were no differences in both the frequencies of the *S2* and *M2* alleles between myocardial infarction survivors and healthy subjects. The alleles identified by the apo CIII *Sst*-I and apo AI *Msp*-I polymorphisms were observed to be in linkage disequilibrium (Δ=0.206±0.059, P<0.001). Three of the four possible haplotypes were identified: the haplotype frequencies were *S1-M1*=0.604, *S2-M2*=0.341 and *S1-M2*=0.055 in healthy subjects. The individual with the haplotype *S1-M2* was significantly increased in myocardial infarction survivors compared with healthy subjects (24% versus 11%; χ^2=4.90, df=1, P<0.05). The data suggest that the haplotype *S1-M2* may be a linkage marker for the putative atherogenic gene in the apo Al-CIII-AIV gene complex in Japanese.

ACKNOWLEDGMENTS

We are grateful to Dr. S. E. Humphries, St. Mary's Hospital Medical School, London for giving us a human apo AI clone. We thank Dr. T. Sakuma, Tsukuba Medical Center for giving us samples from myocardial infarction survivors. We also grateful to Miss Mariko Sugawara, University of Tsukuba for the preparation of this manuscript.

REFERENCES

1. Herbert, P. N., Assman, G., Gotto, A. M. Jr., and Fredrickson, D. S., in "The Metabolic Basis of Inherited Disease" (J. B. Stanbury, J. B. Wyngaarden, D. S. Fredrickson, J. L. Goldstein, and M. S. Brown, eds.), p. 589, McGraw-Hill, New York, 1983.
2. Karathanasis, S. K., McPherson, J., Zannis, V. I., and Breslow, J. L., *Nature* 304:371 (1983).
3. Law, S. W., Gray, G., Brewer, H. B. Jr., Sakaguchi, A. Y., and Naylor, S. L., *Biochem. Biophys. Res. Commun.* 118:934 (1984).
4. Cheung, P., Kao, F. T., Law, M. L., Jones, C., Puck, T. T., and Chan, L., *Proc. Natl. Acad. Sci. USA* 81:508 (1984).
5. Karathanasis, S. K., *Proc. Natl. Acad. Sci. USA* 82:6374 (1985).
6. Elshourbagy, N. A., Walker, D. W., Bogushki, M. S., Gordon, J. I., and Taylor, J. M., *J. Biol. Chem.* 261:1998 (1986).
7. Karathanasis, S. K., Norum, R. A., Zannis, V. I., and Breslow, J. L., *Nature* 301:718 (1983).
8. Karathanasis, S. K., Zannis, V. I., and Breslow, J. L., *Nature* 305:823 (1983).
9. Ferns, G. A. A., Stocks, J., Ritchie, C., and Galton, D. J., *Lancet* II: 300 (1985).
10. Rees, A., Stocks, J., Williams, I. G., Caplin, J. L., Jowett, N. I., Gamm, A. J., and Galton, D. J., *Atherosclerosis* 58:269 (1985).
11. Ferns, G. A. A., and Galton, D. J., *Hum. Genet.* 73:245 (1986).
12. Ordovas, J. M., Schaefer, E. J., Salem, D., Ward, R. H., Glueck, C. J., Vergagni, C., Wilson, P. W. F., and Karathanasis, S. K., *N. Engl. J. Med.* 314:671 (1986).
13. Buraczynska, M., Hanzlik, J., and Grzywa, M., *Hum. Genet.* 74:165 (1986).

14. Sidoli, A., Giudici, G., Soria, M., and Vergani, C., *Atherosclerosis* 62:81 (1985).
15. Rees, A., Stocks, J., Paul, H., Ohuchi, Y., and Galton, D. J., *Hum. Genet.* 72:168 (1986).
16. Kunkel, L. M., Smith, K. D., Boyer, S. H., Borgaonkar, D. S., Wachtel, S. S., Miller, O. J., Berg, W. R., Jones, H. W., and Rary, J. M., *Proc. Natl. Acad. Sci. USA* 74:1245 (1977).
17. Southern, E. M., *J. Mol. Biol.* 98:503 (1975).
18. Kessling, A. M., Horsthemke, B., and Humphries, S. E., *Clin. Genet.* 28:296 (1985).
19. Onuki, M., Iwamura, Y., Humphries, S. E., Satoh, J., Hattori, N., Yamakawa, K., Yamanouchi, Y., Okafuji, T., Tsuchiya, S., and Hamaguchi, H., *Jpn. J. Hum. Genet.* 31:(in press).
20. Satoh, J., Hattori, N., Onuki, M., Yamakawa, K., Fujiwara, H., Amamiya, H., Nagaoka, H., Sakuma, T., Yamanouchi, Y., Okafuji, T., Iwamura, Y., Tsuchiya, S., Fukutomi, H., Ohsuga, T., and Hamaguchi, H., *Jpn. J. Hum. Genet.* 32:(in press).
21. Seilhamer, J. T., Protter, A. A., Frossard, P., Levy-Wilson, B., *DNA* 3:309 (1984).
22. Rees, A., Stocks, J., Sharpe, C. R., Vella, M. A., Shoulders, C. C., Katz, J., Jowett, N. I., Baralle, F. E., and Galton, D. J., *J. Clin. Invest.* 76:1090 (1985).

RFLPS IN THE INSULIN RECEPTOR GENE AND TYPE 2 DIABETES IN THE PACIFIC[1]

S.W. Serjeantson
B.S. White

Department of Human Genetics
John Curtin School of Medical Research
Australian National University
Canberra, Australia

G.I. Bell

Chiron Corporation
Emeryville, California, U.S.A.

P. Zimmet

Lions-International Diabetes Institute
Royal Southern Memorial Hospital
Caulfield South
Victoria, Australia

I. INTRODUCTION

Genetic factors play a major role in predisposition to Type 2, or non-insulin-dependent, diabetes mellitus in Pacific populations (Serjeantson and Zimmet, 1984) as elsewhere (Neel, 1976). Pedigree analysis in Micronesians has shown that most of the variation in glucose tolerance can be attributed to a major autosomal gene with a dominant mode of inheritance (Serjeantson et al., 1986). This gene is clearly not linked with the major histocompatibility complex on Chromosome 6, nor with DNA restriction fragment length polymorphisms (RFLPs) associated with the insulin gene on Chromosome 11 (Serjeantson et al., 1983). The recent cloning of cDNA for the insulin receptor gene (Bell, unpublished; Ullrich et al., 1985; Ebina et al., 1985) has provided the

[1]Supported by NIH grant AM25446

opportunity to examine the role of this gene in the pathogenesis of diabetes. Here, we describe the search for RFLPs in the insulin receptor gene in Micronesians and examine the distribution of these in diabetic patients and their healthy relatives.

II. MATERIALS AND METHODS

A. Study Population

Adult Micronesians from the Western Pacific Republic of Nauru were examined for glucose tolerance when they presented with informed consent at a general medical survey (Zimmet et al., 1984). Plasma glucose concentrations were measured following overnight fasting and two hours after a 75 g oral glucose load, using methods described elsewhere (Zimmet *et al*., 1977). Eight extended families with histories positive for Type 2 diabetes were selected for further and more recent study. Two of these families were of mixed Micronesian and European ancestry. Diabetes patients and their healthy adult relatives provided 40 ml of heparinized blood. Plasma, buffy coats and red blood cells were separated within 12 hours of collection and shipped in dry ice to Canberra, Australia, where DNA was extracted from buffy coats from a total of 170 volunteers. For families with offspring aged between 20-29 years, the prepared DNA was used for confirmation of Mendelian segregation of RFLPs. These families were not included in linkage studies, since young adults with normal glucose tolerance may yet develop Type 2 diabetes in maturity.

B. Screening for Polymorphisms

Genomic DNA was prepared from buffy coats using standard methods (Maniatis et al., 1982). Approximately 10 μg of DNA from 18 diabetes patients and 21 healthy individuals was digested with *Taq* I, *Bgl* II, *Msp* I, *Rsa* I, *Bam* HI, *Eco* RV, *Stu* I, *Pst* I, *Pvu* II and *Hind* III. Fragments were separated by horizontal electrophoresis through 0.8% agarose gels for 16 hours then transfered to Gene Screen Plus membrane according to manufacturer's (New England Nuclear) specifications. DNA fragments were hybridized with two insulin receptor probes, HIR13.1 and HIR18.2. These clones encode all but the 5'- untranslated region of the human insulin receptor cDNA.

RFLPs were observed when Bgl II and Rsa I DNA fragments were hybridized with HIR13.1. These RFLPs were confirmed by repeat digests and by clear segregation of fragments in families. For linkage analysis Bgl II results were available for 68 individuals and Rsa I results for 64.

C. Linkage Analysis

Linkage analysis was performed for Type 2 diabetes and insulin receptor gene RFLPs. Linkage is measured by the maximum likelihood estimate of the recombination fraction (θ), based on the relative probability (Pr) of having generated the observed family distribution. Then Log_{10} Pr is defined as the lod score. Lod scores were calculated using the computer program LIPED (Ott, 1974). Although evidence favours a dominant mode of inheritance of Type 2 diabetes in Micronesians (Serjeantson and Zimmet, 1984), linkage analysis was performed with Type 2 diabetes considered as a dominant gene with complete penetrance or as a recessive trait with 90% penetrance. Under the recessive model it was necessary to prescribe less than 100% penetrance due to the occurrence of several non-diabetic individuals with both parents diabetic.

III RESULTS

A. Insulin Receptor Gene Polymorphisms in Micronesians.

Probe HIR13.1 hybridized with numerous Taq I and Msp I bands in Micronesians, but these were invariant. For example, Taq I generated fragments of 5.4, 4.1, 3.6, 2.6, 2.2, 2.0, 1.7, 1.6 and 1.1 kb, but these same fragments were present in all individuals tested. Msp I generated bands of 5.0, 4.6, 4.1, 2.8, 2.4, 2.2, 1.9, 1.4, 1.3 kb and although the occasional sample was missing the 2.4 kb band, this rare variant was not pursued further. Similarly, possible variants were seen in Pst I and Hind III digests with probe HIR18.2, but since these variants were rare, they were not examined further in family material. The more commonly observed RFLPs occurred when Bgl II and Rsa I digests were hybridized with HIR13.1.

The Bgl II polymorphism showed two allelic fragments of sizes 23 kb and 20 kb. These fragments clearly segregated in families and homozygotes were observed for both alleles. The Rsa I RFLP showed two allelic fragments of 6.8 kb and 6.2 kb.

b. Insulin Receptor RFLPs in Diabetes Patients and Controls

Table I shows the genotype distributions of the insulin receptor RFLPs in diabetes patients and controls. For this series, ten diabetes patients were matched with ten healthy sibs. An additional 8 unrelated diabetes patients and 11 unrelated healthy individuals were included to provide random population distributions of the insulin receptor RFLPs.

TABLE I. Distribution of Insulin Receptor RFLPs in Diabetes Patients and Controls

Genotype		Patients (N=18)		Controls (N=21)	
Enzyme	Alleles	N	%	N	%
Bgl II	23,23 kb	6	33.3	8	38.1
	23,20	6	33.3	8	38.1
	20,20	6	33.3	5	23.8
Rsa I	6.8,6.8	5	27.8	5	23.8
	6.8,6.2	6	33.3	6	28.6
	6.2,6.2	7	38.9	10	47.6

There are no differences in genotype distributions in diabetes patients and controls, for either the *Bgl* II or the *Rsa* I RFLPs. In the control group, the frequency of the 23 kb *Bgl* II fragment is 0.444 and the frequency of the 6.8 kb *Rsa* I fragment is 0.381. The two polymorphic sites generate nine possible genotypes for the insulin receptor gene, as shown in Table 2. Once again, although the numbers in each category are small, no differences are observed in genotype distributions in diabetes patients and controls. Table 2 shows that all genotypic combinations are represented in the population, suggesting absence of strong linkage disequilibrium between the two polymorphic sites. In the control series, there is completely random association between the alleles. However, in diabetes patients there is a tendency (X_1^2 = 2.92) for the 23 kb *Bgl* II and the 6.8 kb *RsaI* fragments to occur cojointly; similarly the 20 kb *Bgl* II and 6.2 kb *Rsa* I fragments co-occur.

The absence of strong linkage disequilibrium within the insulin receptor gene suggests that a strong association between diabetes and RFLPs is unlikely to be detected in a random population series, even if the insulin receptor gene itself was implicated in Type 2 diabetes. Therefore linkage analysis is more useful for examining the possible role of the insulin receptor gene in diabetes.

TABLE II. Distribution of Insulin Receptor Genotypes in Diabetes Patients and Controls.

Genotype				Patients	Controls
Bgl II		Rsa I		(N=18)	(N=21)
23kb	20kb	6.8kb	6.2kb	Number	Number
+	+	+	+	3	5
+	+	+	-	1	1
+	+	-	+	2	2
+	-	+	+	2	1
+	-	+	-	3	4
+	-	-	+	1	3
-	+	+	+	1	4
-	+	+	-	1	0
-	+	-	+	4	1

C. Linkage Analyses

Lod scores are given in Table III for linkage between insulin receptor gene RFLPs and Type 2 diabetes. Values of the recombination fraction (θ) range from 0.0 to 0.4 and in Table III, θ in males is set equal to that in females. Lod scores were calculated also for linkage between the Bgl II and Rsa I sites. Although these polymorphic sites fall within a single gene, the maximum lod score obtained was +0.74 at θ=0.1. When the value of θ was varied in males and females, the maximum value occurred once again when θ =0.10 in both males and females.

TABLE III. Lod Scores for Linkage of Type 2 Diabetes (DM) with Insulin Receptor Gene RFLPs

Genetic markers	Recombination fraction (θ) 0.0	0.1	0.2	0.3	0.4
Bgl II, Rsa I	0.70	0.74	0.58	0.34	0.11
DM dominant					
DM, Bgl II	-0.34	-0.63	-0.29	-0.12	-0.03
DM, Rsa I	0.10	-0.34	-0.12	-0.04	-0.01
DM recessive					
DM, Bgl II	-0.99	-0.52	-0.26	-0.10	-0.02
DM, Rsa I	-0.48	-0.23	-0.11	-0.04	-0.01

As shown in Table III, linkage of Type 2 diabetes with the *Bgl* II or *Rsa* I polymorphic sites of the insulin receptor gene could not be demonstrated, irrespective of whether diabetes was considered as a dominant trait with complete penetrance or as a recessive trait with 90% penetrance. However, since lod scores did not fall below -2, the possibility of linkage cannot be definitely excluded.

IV DISCUSSION

The absence of strong linkage disequilibrium between the *Bgl* II and *Rsa* I polymorphic sites of the insulin receptor gene suggests that even if this gene plays a role in Type 2 diabetes, population associations between RFLPs and diabetes may be quite weak. Random associations between the *Bgl* II and *Rsa* I markers have been reported also in American Blacks, Pima Indians and in Caucasoids (Elbein et al., 1985). In these populations, the frequency of the *Bgl* II 20kb fragment was 0.26 in Blacks, 0.02 in the Pima and 0.17 in Caucasoids, in comparison with 0.43 in Micronesians. The allelic frequency of the *Rsa* I 6.2 kb fragment was 0.67 in Blacks, 0.17 in Pima and 0.48 in Caucasoids, compared with 0.62 in Micronesians. The absence of strong linkage disequilibrium in this distal region of the short arm of Chromosome 19 raises additional questions regarding the genetics of the insulin receptor. On the basis of the size of the bands seen on Southern blot analysis, the coding region of the insulin receptor gene spans more than 40 kb, although the cDNA is approximately 5 kb. Ullrich et al., (1985) have suggested that there is only one copy of the insulin receptor gene, on the basis of Southern blot analysis with a cDNA fragment from the 3' - most terminal untranslated sequence. However, more than one mRNA has been identified (Ullrich et al., 1985). The lack of close linkage between the two polymorphic sites of the insulin receptor gene could arise if there was duplication of the gene on Chromosome 19, with sufficient homology to cross-hybridize with a cDNA probe but with base sequence divergence in the 3'- untranslated region. An alternative explanation is that a recombinational hot spot occurs in the region of the insulin receptor gene. Certainly, it is becoming increasingly evident that recombination is non-random in nature (Barker et al., 1984). For instance, recombination within a 9kb interval 5' to the β-globin gene on Chromosome 11 is exceedingly common (Kazazian et al., 1984), whereas recombination is reduced at the serum albumin locus (Murray et al, 1984), probably due to its proximity to the centromere on Chromosome 4.

Since population associations between Type 2 diabetes and insulin receptor gene RFLPs were weak, we examined joint segregation of diabetes and RFLPs in eight extended families. These analyses did not provide any evidence that the insulin receptor gene is involved in the pathogenesis of Type 2 diabetes but further families must be studied to definitely exclude a putative role for the gene in this disease. Southern blot analyses showing normal patterns of DNA in diabetes patients do not, of course, exclude the possibility that a defective messenger RNA may be involved and studies of possible transcriptional anomalies in diabetes patients have yet to be undertaken.

RFLPs associated with the insulin receptor gene, as with the insulin gene, appear not to be related to Type 2 diabetes in the Pacific. The search for the susceptibility locus may now need to concentrate on random DNA markers distributed throughout the genome.

REFERENCES

Barker, D., Holm, J., and White, R. (1985). Am.J.Hum. Genet. 36:1159.

Ebina, Y., Ellis, L., Jarnagin, K., Edevy, M., Graf, L. et al. (1985). Cell 40:747.

Elbein, S.C., Corsetti, L., Ullrich, A., and Permutt, M.A. (1986). Proc.Natl.Acad.Sci 83:5223.

Kazazian, H.H., Orkin, S.H., Markham, A.F., Chapman, C.R., Youssoufian, H., and Waber, P.G. (1984). Nature 310:152.

Maniatis, T., Fritsch, E.F., Sambrook, J. (1982). "Molecular Cloning. A Laboratory Manual." Cold Spring harbour, Laboratory, Cold Spring Harbor, New York.

Murray J.C., Mills, K.A., Demopulos, C.M., Hornung, S., and Motulsky, A.G. (1984). Proc.Natl.Acad.Sci 81:3486.

Neel, J.V. (1976), in "The Genetics of Diabetes Mellitus" (W. Creuzfeldt, J. Kobberling and J.V. Neel, eds.) p.1. Springer Verlag, Berlin.

Ott, J. (1974). Am.J.Hum.Genet 26:588.

Serjeantson, S.W., Owerbach, D., Zimmet, P., Nerup, J., and Thoma, K. (1983). Diabetologia 25:13.

Serjeantson, S., and Zimmet, P. (1984) In: "Diabetes Mellitus: Recent Knowledge on Aetiology, Complications and Treatment". (S.Baba, M.Gould, and P.Zimmet, eds.) p.23. Academic Press, Sydney.

Serjeantson, S.W., Zimmet, P., and Morton, N.E. (1986).

Genetic Epidemiology (in the press).
Ullrich, A., Bell, J.R., Chen, E.Y., Herrera, R., Petruzzeli, L.M. et al. (1985). Nature 313:756.
Zimmet, P., King, H., Taylor, R., Raper, L.R., Balkau, B., Borger, J., Heriot, W., and Thoma, K. (1984). Diabetes Res. 1:13.
Zimmet, P., Taft, P., Guinea, A., Guthrie, W., and Thoma, K. (1977). Diabetologia 13:111.

DEMONSTRATION OF MAJOR GENE FOR CEDAR POLLINOSIS

Norikazu Yasuda[1]

Division of Genetics
National Institute of Radiological Sciences
Chiba, Japan

Masahiko Muto

Department of Clinical Genetics
Medical Institute of Bioregulation
Kyushu University
Oita, Japan

Yozo Saito

Department of Otorhinolaryngology
School of Medicine
Tokyo Medical and Dental University
Tokyo, Japan

Takehiko Sasazuki

Department of Genetics
Medical Institute of Bioregulation
Kyushu University
Fukuoka, Japan

[1]Supported in part by a grant from Ministry of Health and Welfare, Japan.

I. INTRODUCTION

In common diseases a relatively small numbers of potentially identifiable major genes may contribute to the genetic etiology and explain most of the genetic variation. Such genes do not act in vacuum. The ensemble of all other genes against which such major genes act are considered to be the various polygenes that constitute the genetic background. It is well known that the genetic background may modify and influence expression of major genes(1).

In order to demonstrate such major genes, Cedar pollinosis has been selected for an illustration of statistical strategy, since Cedar pollinosis allergen is known from the environmental origin, and host response to the allergen is simply taken as genetic factors. This is especially crucial advantage in analysis of Cedar pollinosis because in the pathogenesis of most common diseases we do not know the environmental agents which interact with the genetic factors to develop the disease.

In this communication, we present a statistical analysis of 165 families with Cedar pollinosis, specifically Cryptomeria pollinosis. We will show that the susceptibility to Cedar pollen antigen is in main controlled by a recessive gene linked to HLA with reduced penetrance to ca 50 percent, and that this major gene could be in linkage disequilibrium with HLA-DQw3 gene.

II. MATERIALS AND METHODS

One hundred and sixty-five families with Cedar pollinosis were gathered through 107 probands which consisted of 30 males and 77 females. The patients with Cedar pollinosis were sampled according to the following diagnostic criteria. (i) Clinical history during pollination season, usually from February to April in Japan; (ii) Positive skin test by crude Cedar pollen extract; (iii) Positive histological examination of nasal discharge; and (iv) Positive radioallergosorbent test. Family members younger than five years old were excluded from the genetic analysis, because Cedar pollinosis is extremely rare below this age. In all 766 family members served as the subjects of this study. Sex ratio of probands was 2.6 females to 1 male. But, since the corresponding figure in the first degree relatives was 1 to 0.9, no sex bias was observed in manifestation of Cedar pollinosis.

HLA-A, -B and -C typings were performed by the NIH microcytotoxicity method using 110 well-defined HLA typing sera. HLA-DR typing was made by the method used in the Eighth International Histocompatibility Workshop.

Out of 165 families, 85 families whose both parents are normal and 3 families of normal x affected backcross type were selected through proband, namely, the ascertainment was single with the probability $\pi \to 0$. Five normal x normal matings, 68 backcross types and 4 families of both parents affected were ascertained through parent so that the ascertainment probability was $\pi=1$.

III. RESULTS

Two simple genetic hypotheses on the major gene for the susceptibility to Cedar pollinosis have been tested. Dominant inheritance could be ruled out from the family data. There were too many affected children segregated from matings of both normal parents. Furthermore, we observed four families in which all family members were affected. These observations show rather to be recessive mode of inheritance.

Table I. Segregation Analysis of Cedar Pollinosis for Testing Recessive Hypothesis

Genetic hypothesis		Mating type	No of family	χ^2 for a goodness of fit test(df)	P
Simplex +	$p=\frac{1}{4}$, $x=0$, $\pi \to 0$	N x N	84	31.81(13)	<.005
Multiplex	$\hat{p}=.12 \pm .02$, $x=0$, $\pi \to 0$	N x N	84	9.73(12)	>.06
Multiplex	$p=\frac{1}{4}$, $\pi \to 0$	N x N	22	2.64(6)	>.80
	$\hat{p}=.11 \pm .07$, $\pi \to 0$	N x N	22	1.09(5)	>.90
	$p=\frac{1}{2}$, $\hat{h}=.64 \pm .09$, $\pi=1$	N x A	52	4.27(6)	>.60

p:expected segregation frequency $\hat{p}$:observed segregation frequency by a maximum likelihood method x:proportion of sporadic cases π:ascertainment probability $\hat{h}$:observed proportion of nonsegregating families
N:Normal A:Affected

The family data were then analyzed under the null hypothesis that the major gene for susceptibility to Cedar pollen antigen was recessive mutant. Two subsets of family data were subjected to segregation analysis(2): (i) eighty-four intercrosses ascertained through children; and (ii) 52 backcrosses ascertained through parent (Table I). In 84 intercrosses the expected segregation frequency was highly significant. When the segregation frequency was estimated to be .12 and adjusted with reduced penetrance of 48 percent (= 100x.12/.25), the significance was disappeared. Among 84 intercross families, 22 couples have had more than one affected child. Segregation analysis applied to those multiplex families yielded the similar segregation frequency (.11) and penetrance (.44) as those obtained from the analysis with multiplex and simplex families. This result suggests that there would be little genetic heterogeneity in Cedar pollinosis among families studied.

Segregation analysis of 52 backcross mating types ascertained through parent showed that a family distribution of Cedar pollinosis was compatible with the recessive hypothesis on the major gene and the gene frequency was calculated as .22 from the proportion of nonsegregating family that was .64.

Out of 16 affected sibpairs observed, 9 shared two HLA haplotypes identical by descent, 7 shared one HLA haplotype identical by descent, and none of affected sibpair was nonidentical. This distribution of HLA haplotypes was significantly different from random expectation (Table II).

Table II. Affected Sibpair Analysis

#	Observed	Random expected	Recessive expected	Dominant expected
Two	9	4	9.18	8
One	7	8	5.88	8
None	0	4	.90	0
χ^2		10.38*	1.16	0.25
df		2	1	1
gene frequency			.32±.10	0

No. of haplotype shared by affected sibpair

This suggests that a gene linked to HLA loci has a major role in determining the susceptibility to Cedar pollen antigen. The minimum chisquare .25 was obtained under a dominant hypothesis on the major gene with the frequency zero, while a recessive mode of inheritance fitted as well. A goodness of fit chisquare was 1.16 with one degree of freedom. The derived frequency of recessive gene was .32. We take a recessive hypothesis on the major gene because zero frequency of dominant gene is biologically untenable and the result of segregation analysis was compatible with the recessiveness. Furthermore, estimated frequency of recessive gene by the affected sibpair method(3) was not significantly different from the one obtained by segregation analysis.

The expected frequency of homozygous individuals of the HLA-linked susceptibility gene in the general population is $(.32)^2=.102$, while the observed incidence among Japanese is .07(4). So the estimated penetrance is 69 percent by the affected sibpair method.

Table III. Lod Score Analysis* for Cedar Pollinosis and HLA Allowing Reduced Penetrance

Penetrance (%)	Recombination fraction (θ)					
	.00	.02	.04	.06	.08	.10
100	-10.34	2.39	3.13	3.37	3.42	3.35
90	2.62	3.04	3.10	3.06	2.96	2.81
80	2.74	2.77	2.72	2.62	2.49	2.35
70	2.52	2.45	2.35	2.24	2.11	1.98
60	2.26	2.17	2.09	1.95	1.83	1.71
50	2.06	1.95	1.85	1.74	1.62	1.51

* 19 informative nuclear families were subjected to the analysis.

The lod scores were calculated from 19 HLA typed informative families(5) (Table III). Under the hypothesis of complete recessive mode of inheritance, the maximum lod score was 3.42 at the recombination fraction θ=.08. We observed however reduced penetrance of 48 or 69 percent. Implementing reduced penetrance in linkage analysis(6), lod scores for a specific value of reduced penetrance ranged from 100 to 50 percent attained a maximum from 2.06 to 2.52 at near zero recombination fraction. Thus the result is favor for evidence to a tight linkage between HLA and a major gene for the susceptibility to Cedar pollinosis. It would be necessary to gather further informative families, perhaps two or three times more, in order to clarify the linkage relationship.

Difference in HLA antigen frequency between patient and control groups were tested by chisquare analysis with Yates' correction. Each P value was corrected by the number of antigen tested. The strength of association was measured by the relative risk calculated by the cross-product ratio in

Table IV. Association between Cedar Pollinosis and HLA[1]

Antigen	Patient(%)	Control(%)	Relative risk
Aw33	19/97(19.6)	17/212(8.0)	2.79
B44	24/97(24.7)	22/220(10.0)	2.96
DRw13	10/43(23.3)	6/68 (8.9)	3.13
DQw3	40/43(93.0)	43/68 (63.2)	7.75*

* Significant at 5 percent level
1) Yates' correction was applied for the statistical test.

2 x 2 table. Significant association was found between Cedar pollinosis and HLA-DQw3. The relative risk was 7.75 and the corrected P value was less than .05 (Table IV). Reconciling the result with one obtained from linkage analysis, the major gene was in linkage disequilibrium with HLA-DQw3 allele among Japanese.

IV. DISCUSSION

Cryptomeria pollinosis is an allergy to the antigen from pollen of <u>Cryptomeria japonica</u> commonly found in Japan. Cryptomeria pollen observed from February to April every year in Japan, except Hokkaido and Okinawa. Segregation analysis and affected sibpair method have suggested that a recessive gene is mainly responsible for the susceptibility to Cedar pollen antigen. We observed however reduced penetrance of 48 or 69 percent. The reduced penetrance observed in the present study might be explained by the presence of an insufficient exposure of population to Cedar antigen, the idea came from the observation that the number of patients with Cedar pollinosis was higher when a high count of Cedar pollen was found. However, another important possibility for the reduced penetrance is that a second major locus or 'a set of polygenes' as modifier may be involved in the development of Cedar pollinosis. The true mode of inheritance might probably not be simple: it may involve other gene(s) not linked to HLA.

Search for 'the second major gene' has attempted by extended segregation analysis for two loci with nuclear family data. Three hypotheses for the mode of inheritance would be of interest: double recessives, double dominants and dominant-recessives. The segregation frequency of affected individuals for a given type of mating is now complex, so named as complex

Table V. Two Locus Segregation Analysis of Cedar Pollinosis

	84 N x N matings ($\pi \to 0$)			52 N x A matings ($\pi = 1$)	
Genetic hypothesis	χ^2 for a goodness of fit (df=13)	(df=12)#	(t)		(df=5)
R (p=1/4)	31.18**	9.73	(.48)	(p=½)	4.27
R-R(p=1/16)	15.12	9.37	(.63)	(p=¼)	4.59
R-D(p=1/8)	26.70*	9.32	(.50)	(p=¼)	5.94
D-D(p=1/4)	37.67**	9.51	(.45)	(p=½)	9.28

R: recessive D: dominant (t): penetrance
Adjusted to penetrance * P<.05 ** P<.01

segregation analysis. We performed complex segregation analysis on these three hypotheses and compared with the single recessives examined before (Table V). The result of analysis shows that the double recessive mode of inheritance seems to be most compatible for the susceptibility to Cedar pollen antigen. But when the segregation frequency was adjusted with reduced penetrance, the development of Cedar pollinosis could be explained by any of four genetic hypotheses.

V. SUMMARY

In summary we propose the following genetic hypothesis for the development of Cedar pollinosis. First person must be exposed to Cedar pollen. Next, one major gene linked to HLA will at least be responsible for the susceptibility to Cedar pollen. Another major gene not linked to HLA might exist in the 'genetic background'. Although the HLA-linked major gene was demonstrated to be recessive, the penetrance was reduced to ca 50 percent by either insufficient exposure to Cedar pollen or the genetic background.

REFERENCES

1. Vogel, E. and Motulsky, A. G. (1979). "Human Genetics: Problems and Approaches." Springer-Verlag, Berlin, p. 186.

2. Morton, N. E. (1956). Am. J. Hum. Genet., 11:1.
3. Thomson, G. and Bodmer, W. F. (1977). In "Measuring Selection in Natural Population" (T. B. Christiansen, T. M. Fenchel and O. Barndorff-Nielsen), p. 545. Springer-Verlag, Berlin.
4. Nomura, K. and Uchikoshi, S. (1981). Jpn. J. Allergol., 30:645. (in Japanese)
5. Morton, N. E. (1955). Am. J. Hum. Genet., 7:277.
6. Ott, J. (1985). "Analysis of Human Linkage." Johns Hopkins University Press, Baltimore.

BIOCHEMISTRY OF HLA CLASS II ANTIGENS: A TOOL FOR GENETIC ANALYSIS OF INTRACTABLE DISEASES

Dominique Charron

Laboratoire d'Immunogenetique Moleculaire
15, rue de l'Ecole de Medecine 75006 Paris, France

Département d'Hématologie CHU Pitié Salpêtrière
91, Bd de l'hopital 75013 Paris, France

I. INTRODUCTION

The involvement of the human leucocyte antigen (HLA) system and particularly of the HLA-D region has represented over the past decade one of the most intensively investigated areas in human immunogenetics.

Molecular biology, biochemical and cellular approaches have provided experimental data which presently allow to propose some explanations for the relation of the HLA system with increase susceptibility to numerous diseases.

Three aspects can be particularly emphasized. 1) Some alleles of the HLA-D region represent indeed genetic markers for disease susceptibility or resistance. 2) Aberrant or increased expression of HLA class II antigens may lead to autoimmunity. 3) Bone marrow and organ transplantation are critically under control of the HLA-D region. Moreover anti-Ia therapy has been recently reported to be successful in animal models to delay or prevent the occurrence of autoimmune disease and is presently investigated in clinical medicine.

A better knowledge of the qualitative and quantitative aspects of HLA class II antigen structure and expression represents an important step in understanding the role of Ia-antigens in physiology and pathology.

The HLA class II antigens are cell surface molecules encoded within the HLA-D region which are primarily involved in the regulation of immune responses by controlling the interactions which take place between macrophages and helper T lymphocytes in antigen presentation and between T and B cells for antigen-specific antibody production (1,2). MHC class II antigens also elucidate the mixed lymphocyte reaction (MLR) *in vitro* and the graft versus host reaction (GVH) *in vivo* (3).

The structural diversity of human Ia antigens arises not only from the genetic complexity of the system which generates a high level of polymorphism in the population but also from regulatory processes and post-translational modifications. A class II gene or set of genes may be differentially expressed in different cells and tissues (4). Since the surface of the immunocompetent cells represents the site of action of the class II products it is of critical importance to consider the different aspects of the structural diversity of the human Ia products in order to unravel the precise mechanisms which underlie their functions. I will review here recent data on the biochemistry of the human MHC class II antigens and emphasize some of the structural features such as the biochemical polymorphism, glycosylations and differential expressions which may be relevant to the function of the Ia antigens.

A. Biochemical polymorphism of the HLA-DR haplotypes

Extensive data on the number and the structure of the genes coding for the class II chains has been obtained over the past years. However it concerns thus far a limited number of haplotypes. The type and number of MHC class II genes expressed in one individual are still largely unknown. For several years we have used biochemistry to investigate the polymorphism of the HLA-D region products (5,6). While at first it was apparent that a unique DR molecule was associated with each serologically defined DR type the situation became more and more complex when additional haplotypes were analyzed. Two dimensional gel electrophoresis (2D-PAGE) allows a detailed characterization of the number and the association of class II molecules expressed in several HLA-DR haplotypes.

DR1 and DR BON haplotypes

One of the simplest HLA-DR haplotypes, HLA-DR1, possess only one DRβ chain expressed (6). No structural variant of this DR β chain is presently known. It is noteworthy that while most of the serological DR types are recognized by monospecific allo antisera (anti-DR3, DR4, etc...) as well as by supertypic ones (anti-DRW52, DRW53) the DR1 specificity is solely defined by monospecific reagents (anti-DR1). No anti-"DRW51" sera have been found and this may reflect that only one DR β chain is expressed in DR1 cells. We have recently described a new biochemical variant, DR BON which may be closely related through evolution with DR1 since it shares two important features with this latter haplotype: DR BON posses also only one expressed DR β chain and DR BON and DR1 cannot be distinguished by restriction fragment length polymorphism (RFLP) analysis with a DR probe (clone 2918.4) using four enzymes (BgI II, Bam HI, Pvu II, Eco R1) (7).

DR2 haplotypes

The HLA-DR molecules were analyzed in several different lymphoblastoid cell lines homozygous at the HLA-D region all typed by serology as DR2. Five DW clusters were identified by cellular typing (mixed lymphocyte reaction) which are designated DW2, DW 12, FJO, DB9, DWX. The structural polymorphism of the DR and DQ molecules expressed in these cells was assessed by (2D PAGE) using non equilibrium pH gradient (NEPHGE) in the first dimension to analyse the β chains and alternatively isoelectrofocalisation (IEF) in order to better resolve the area of the α chain. Each homozygous cell type was shown to possess at least two distinct HLA-DR molecules per haplotype. The DR α and the DR β 1 chains are common to all the DR2 cell lines and the DR β 2 chains are strikingly variable among the different HLA-DW cellular subtypes. Four electrophoretic DR β variants can be identified in DW2, DW12, DWX, DB9 and FJO (8). When the DQβ chains are studied, an extensive structural polymorphism is also detected as well as for the DQ α chains such structural data allow to interpret some of the functional traits observed within the HLA-DR2 haplotypes. The cellular alloreactivity assessed in the mixed lymphocyte reaction is very complex since structural differences in the DR β 2 but also in the DQ β and the DQ α chains may contribute altogether or separately to positive T cell responses. It should be emphasized therefore that any correlation with function can be attained with confidence only when the biochemical analysis of the full set of class II molecules is performed. We are presently performing a biochemical analysis of the class II molecules expressed in juvenile diabetes, multiple sclerosis and narcolepsia diseases which are known to be negatively or positively associated with DR2 in order to detect which are the variants of the DR2 molecules present.

DR3 haplotypes

The HLA-DR3 molecules are the most basic species as detected by 2D gel electrophoresis. The HLA-DR3 haplotypes can be split into two variants, the molecular characterization of which await completion. In addition, a DRW52-bearing DR β molecule is usually detected in association with the DR3 β chain although its expression is usually weaker than the DR3 β chain.

<u>DRW6 (DRW13 + DRW14) haplotypes</u>

The DRW6 specificity which in the new nomenclature includes DRW13 and DRW14 is unique since no monospecific allo antisera identifies all the DRW6. Most of the DRW6 serology relies on positive reactions with supertypic allo antisera of individuals which are not recognized by the monospecific allo antisera used to define the non-DRW6 HLA-DR specificities included in the supertypic group. For example an individual is typed HLA-DRW6 if he is positively recognized by an anti-DRW52 serum while negative with anti-DR3 and anti-DR5 allo antisera. [The DRW52 specificity includes individuals typed as DR3, DR5, and DR6 individuals.] We investigated the DRW6 products in forty distinct individuals and in eight informative families using the segregation of the DRβ chains to identify the haplotype of origin of a given DRβ chain (9,10). Seven distinct DRβ chains were found (DR6-B 1 to DR6-B 7) none of which was common to all the DRW6 individuals. The number of expressed DR β chains varies in different haplotypes. Moreover, when two DR β chains are expressed per haplotype, one may be common to several haplotypes leading to the generation of nine molecular subgroups. These biochemical results may explain the difficulty of the serology. The absence of a common DRβ chain is coherent with the absence of monospecific anti-DRW6 all antisera. In contrast, the biochemical definition of previously unknown subgroups provides a way to select sera specific for these different subtypes. Furthermore since in the population the DRβ 1 chain is not constantly associated with the same DRβ 2 chain it allows to localize to one of the DR β chain functional traits. We have recently shown that the DRW6B5 DR β chain restricts the proliferative response of a DRW6 restricted influenza specific T cell clone (11,12). Moreover a correlation was found between the different DRβ chains and the cellular subtypes of the DRW6 haplotypes. The 6B1 chain defines the DW9 specificity while the association of 6B3 + 6B5 is specific of DW18 and 6B4 + 6B5 of DW19. In the latter cases it is the 6B3 and the 6B4 which bear the allo-reactive determinants which distinguish between DW18 and DW19 since the 6B5 chain is common to the two subtypes. The description and definition of biochemical variants previously undetected by serology suggests to reevaluate the HLA and disease association studies according to the biochemical clusters which correspond to a more exact and complete appreciation of which are the expressed and therefore functionally relevant molecules in a given individual. Since in a recent study DRW6 has been reported to behave as protective in kidney transplantation (13) it would be of considerable interest to investigate in these cases the class II molecules present in the donors and recipients in order to assign a "suppressive potential" to some DRW6 subtypes.

B. Biochemical polymorphism of HLA-DQ molecules

The HLA-DQ antigens have some features which are of special interest. 1) While by serology the DQ specificity appears at first to be supertypic (i.e. DQ anti-sera would recognize several DR haplotypes) the DQ molecules present within one such DQ supertypic group are all structurally different. This strongly supports the idea that the epitope recognized by supertypic allo antisera (or monoclonal anti-DQ antibodies) is shared between molecules which are distinct. This would suggest that the HLA-DQ genes may have been the recipients of extensive but somehow recent gene conversion events which serological signature is reflected in the HLA-DQ supertypic sera. 2) Both the DQ α and the DQ β chains are structurally polymorphic and this enables the formation of hybrid molecules (14). We investigated the possibility of a trans-complementation between a DQ α chain from one parental haplotype and a DQ β chain from the other parental haplotype. Using a monoclonal antibody directed specifically at the DQW2 β chain, we were able to demonstrate that in a DQW1/DQW2 heterozygous cell line both the DQW1 α and the DQW2 α chains were present in the immunoprecipitate (15). These results have been extended to other combinations of haplotypes. Thus transcomplementation occurs within the HLA-DQ subregion and generates novel DQ α , DQ β complexes. A recent report describes an antigen-specific T cell clone obtained from an heterozygous individual which functions only when the antigen is presented by accessory cells carrying both HLA-D region antigens possessed by the donor (16). Such a T cell clone could represent the functional counterpart of the hybrid DQ molecules. We have thus demonstrated that in humans additional polymorphism arises from combinatorial association of class II α - β chains. Mixed haplotype molecules contribute to the qualitative diversity of human class II determinants. Moreover these results may have clinical and physiopathological significance. Susceptibility to insulin-dependent diabetes mellitus (IDDM) is very strongly associated with DR3 and DR4 but the relative risk is even stronger in DR3/4 individuals. This extraordinary heterozygote effect could be due to heterozygous determinants formed by particular combinations of DQα and DQβ chains which in turn would affect the level or the quality of the immune responses and induce autoimmune pathology (17). 3) The level of expression of HLA-DQ molecules is highly variable in normal as well as in abnormal cells. All together these results provide molecular evidence for the existence of a larger repertoire of Ia determinants than was originally thought. The structural complexity described here has important implications for transplantation and HLA and disease association studies. Moreover, the frequent observation that in two distinct haplotypes on DRβ chain can be shared which the second DR β differs allow to consider that linkage disequilibrium between HLA class II genes is weaker than it was assumed previously. 2D PAGE of class II antigens is presently the most accurate method to detect structural variants at the level where they are functional: the cell surface and this biochemical approach should be of great benefit in the future to clinical medicine.

C. Oligosaccharides of HLA class II antigens

Among the post-translational modifications, glycosylations appear to be a major event which affects the final configuration of the cell surface expressed class II molecules. Treatment with tunicamycin and endoglycosidases had previously demonstrated that the class II α chains contain two N-linked oligosaccharides and the β chains only one N-linked carbohydrate moiety. Very little is known on the fine structure of the HLA-DR oligosaccharides (OS). We have recently investigated the oligosaccharide content of the DR antigens using the differential ability of the different OS to bind to distinct lectins. Lectin affinity chromatography analysis has revealed several important points: 1) The DR α chain contains two types of OS: the α 1 fraction has the highest ^{3}H mannose content and consists primarily of high mannose type OS, while the α 2 fraction has only a few high mannose OS. The DR β chain OS are similar to the one observed in the α 2 chain. This data demonstrates the existence of OS heterogeneity of the mature DR α chains. 2) When the OS profiles of HLA-DR molecules are compared in a B lymphoblastoid cell line versus a monocytic cell line striking differences are found with a higher amount of bi-antennary structures and a higher fraction of glycopeptides with no affinity for the ConA lectin in the monocytic line. 3) The OS content of HLA-DR antigens isolated from a B cell line and from a murine L cell (TF229) transfected with DRα and DR β genes and compared the glycosylation patterns with the one of the whole cell membranes. Important differences were found for the OS isolated from the B cell and the L cell TF229. The glycosylation differences were consistent with the pattern observed for the whole membrane and leads to the conclusion that the HLA-DR glycosylation parallels the overall glycosylation of the cell membrane and thus differ from one cell type to another and from one state of activation to another (18,19). These data suggest a great heterogeneity of glycosylation of the class II antigens in different cell types which may affect the cellular recognition events involved in immune responses as it has been suggested in the mouse Ia system.

D. Expression of HLA class II molecules

In contrast to the HLA class I antigens, the HLA class II molecules have a more restricted tissue distribution (20). Some immune functions may require both a specialized cell and the class II expression. Furthermore DR, DQ and DP molecules have discordant expression in different cell populations (4). Class II gene products can thus be considered as tissue differentiation or activation antigens apart from their immunological functions. Since the complex pattern of class II expression has been recently reviewed (4), I will only consider here particular aspects of class II expression. It is apparent from biosynthesis studies that at least two distinct situations are observed concerning the presence of class II antigens at the surface of cells. The B cells and a major fraction of the macrophages express constitutively class II antigens without the need of extrinsic stimuli. This statement can be extended to the EBV-transformed B cell lines or leukemic cells of the B lineage or of monocytic origin. In contrast many cell populations belonging or not to the immune system may express class II antigens only when they are activated. This was demonstrated first in alloreactive T cells (21) and after confirmed in T cell lines and clones (22). Moreover the Ia molecules isolated from activated T cells were shown to be structurally identical to the I a molecules isolated from the B cells of the same individual (21). In disease situations primarily of autoimmune origin, a percentage of circulating T cells may express class II antigens. The precise function of these class II positive T cells and their relevance to disease states is presently unknown. No correlation has been so far found between the number of Ia positive T cells and the severity of the disease or evolution. This may reflect the facts that quantitative tools were lacking to precisely evaluate the level of class

II expression. We have recently developed a Dot blot protein assay which allows to analyze the overall amount of class II antigens present in cells, tissues and fluids (23). This technique may be more able to assess the significance of class II expression in transplantation and autoimmune diseases. Furthermore, defects in class II gene expression have been documented in the lymphocytes of combined immunodeficiency syndrome (24). These patients provide a unique opportunity to investigate some of the mechanisms which regulate Ia expression. Apart from the lymphoid lineage class II antigens may also be present in cells of non-lymphoid origin in both physiological and pathological states. We have recently investigated two such situations. While thymic epithelial cells are class II positive in vivo they rapidly become negative when cultured in vitro (25). A similar situation is found for the synovial lining cells in rheumatoid arthritis (26,27). Nevertheless the class II antigens can be reexpressed in both models when γ interferon is added back to the in vitro culture system suggesting that it may represent at least part of the physiological signal for class II expression in cells not constitutively expressing these antigens (27). The differential expression of class II DR, DQ, and DP antigens on some cell types such as monocytes leukemic cells (28) and activated T lymphocytes would suggest that these gene products have subtle differences in their functions.

E. Concluding remarks

The structural nature of the HLA class II antigens and the level of expression of the various HLA-D region products in different cell types and tissues are critical parameters which regulate the level and quality of immune responses in physiology as well as in pathology. A precise knowledge of the fine structure of the class II products is essential in elucidating the relation of the HLA system with numerous diseases. However important aspects of the cellular biochemistry of the HLA class II physiology are still largely unknown. For example, does postranslational modification of the class II molecules (glycosylation, acylation, phosphorylation) affect their function? How do class II associated molecules (invariant chain, proteoglycans) intervene in the immune response? Answers to these questions are likely to identify biochemical parameters which affect the fine tuning of the immune responses in diseases associated with the HLA system.

BIOCHEMICAL DIVERSITY OF HLA-DRw 6

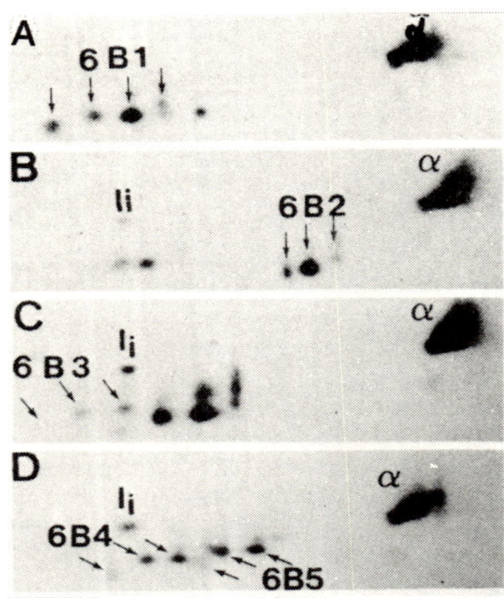

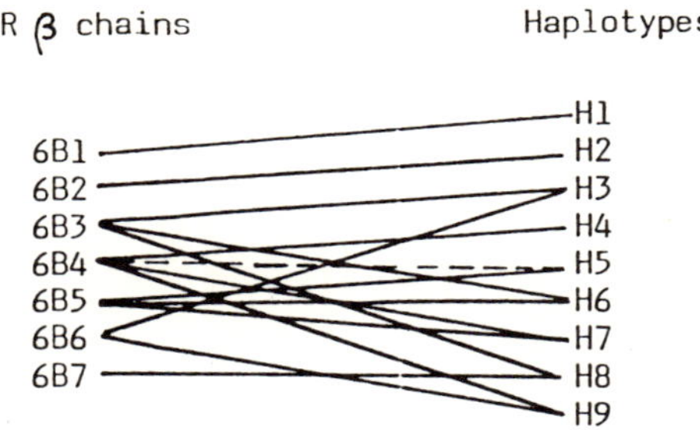

Diagramatic representation of the haplotypic diversity

Correlations between biochemical, cellular, and serologic results

DRw6 Haplotypes	MLR Typing	Serologic Typing
H1	Dw9	DRw14
H2		DRw13
H3		
H4		
H5		
H6	Dw18	
H7	Dw19 or Dw blank	

ACKNOWLEDGMENTS

I thank M. BRANDEL for typing the manuscript.
This work was supported by grants from INSERM, CNRS, ARC, LNCC, ARSEP and FRM to DJC.

REFERENCES

1. Mc DEVITT H.O. In "The rôle of the products of the histocompatibility gene complex in immune responses" (Katz D.H. and Benacerraf B. ed.) 257, Academic, New York (1976).
2. Giles, R.C. and Capra R.C., Adv. in Immunol. 37 : 71 (1985).
3. Bach F.H. Ann. Rev. Genet. 10: 319 (1976)
4. Radkas, Charron D.J., Brodsky F.M., Human Immunol. 16: 390 (1986)
5. Charron D.J. and Mc Devitt H.O. Proc. Natl. Acad. Sci. USA 76 : 6567 (1979)
6. Charron D.J., McDevitt H.O. J. Exp. Med.152: 185 (1980)
7. Coppin H., Villedieu P., Thomsen M., Giraud P., Cambon-Thomsen A. and Charron D. Immunogenetics in press
8. David V., De Roquefeuil S., Freidel C., Gebuhrer L. Lepage V., Betuel H., Debre P. and Charron D.J. Human Immunol. 15, (1986)
9. Haziot A., Lepage V., Degos L. and Charron D.J. In Histocompatibility Testing (Albert et al. ed) Springer Verlag Berlin Heidelberg. 522, (1984)
10. Haziot A., Lepage V., Freidel C., Betuel H. and Charron D. J. of Immunol.
11. Haziot A., Michon J., Freidel C., Degos L., Levy J.P. and Charron D.J. Human Immunology 14: 161 (1985)
12. Haziot A., Michon J., Sterkers G., Degos L., Levy J.P. and D.J. Charron. J. of Immunol. in press (1986)
13. Hendricks G.F.J., Claas F.H.J., Persijn G.G., Witvliet M.D., Baldwin W. and Van Rood J.J. Transplant. Proc. 15 : 1136 (1983)
14. Charron D.J., Lotteau V., Turmel P. Nature 312: 157 (1984)
15. Charron D.J., Lotteau V., Turmel P. In histocompatibility Testing 1984. Eds Albert E. and Mayr W. Springer Verlag Berlin Heidelberg 539 (1984)
16. Hansen G.S., Svejgaard A., Claesson M.H. J. Immunol. 128: 2497(1982)
17. Charron D.J. Path. Biol. 34 : 795 (1986)
18. Nell D., Giner M., Merlu B., Turmel P., Goussault Y. and Charron D.J. Mol. Immunol. 22 : 1053 (1985)

19. Neel D., Merlu B., Mach B., Goussault Y and Charron D.J. (1986)(in press)
20. Charron D.J. Immunol. Communications 10 : 293 (1981)
21. Charron D.J., Engleman E.G., Benike C. and McDevitt H. J. Exp. Med. 152 : 127s (1980)
22. Triebel F., deRoquefeuil S., Blanc C., Charron D.J., Debre P. Hum. Immunol. 15: 302 (1986)
23. Teyton L. Lotteau V., Boyer B., and Charron D.J. J. Immunol. Methods 89 : 73 (1986)
24. Grospierre B., Charron D.J., Durandy A., Griscelli C., Mach B. J. Clin. Invest. 76 : 381 (1985)
25. Berrih S., Arenzana Seisdedos F., Cohen S., Devos R., Charron D.J., Virelizier J.L. J. Immunol. 135: 1165 (1985)
26. Teyton L., Lotteau V., Arenzana Seisdedos F., Pujol, Loyau, Virelizier J.L. and Charron D.J. J. of Immunol. in press
27. Faille A., Turmel P. and Charron D.J. Blood 64 : 33 (1984)
28. Teyton L., Lotteau V., Turmel P., Pujol J.P., Loyau G. Piatier-Tonneau D., Auffray C. and Charron D.J. J. of Immunol.(in press)

II. IMMUNOGENETIC APPROACH

INSULIN RELATED GENES AND DIABETES MELLITUS

Yasunori Kanazawa, Masato Kasuga,
Yoshikazu Shibasaki, Takuya Awata,
Fumimaro Takaku

The 3rd department of Internal Medicine
Faculty of Medicine, University of Tokyo
Tokyo JAPAN

I. INTRODUCTION

Diabetes mellitus is defined as general metabolic disorders due to a deficiency of insulin action. The deficiency is induced either by insufficient secretion of insulin or by insensitivity of the liver and/or other peripheral organs to insulin, or by both. Many known and unknown causes underlie these pathogeneses. The sequences of events initiated by different causes converge to the state of deficient insulin action and can induce the diabetic syndrome.

Several causes of diabetes have already been identified. However, in a majority of the diabetic patients the real cause(s) of the disease are not yet defined. Here, we would like to discuss the genetic approach to diabetes mellitus. However diabetes mellitus is not a disease but a syndrome, and therefore genetically very complicated. Diabetes is truly a nightmare of geneticist (1). Recent advances in the knowledge of human genes have elucidated several diabetes-related genes. They are major histocompatibility genes (HLA), insulin genes, and insulin receptor genes. HLA genes have been extensively studied in relation to diabetes, especially insulin-dependent diabetes, and is discussed elsewhere in this book (2). The insulin receptor gene has been isolated recently and investigated in relation with the diabetic syndrome due to possible receptor abnormalities. This subject is again discussed in elsewhere in this book (3).

Therefore, our discussion in this article is focused on the insulin gene and its related area. The mutation of the insulin gene and the regulatory area of insulin gene expression in relation to diabetes are the major concerns of this paper.

II. STRUCTURE OF THE HUMAN INSULIN GENE AREA

The insulin gene resides on the short arm of chromosome 11 (4) near to lactic dehydrogenase A and the chymotrypsin coding region. The whole structure of the gene was analysed and the sequence of deoxyribonucleic acids was determined (5). The human insulin gene has a length of 1430 base pairs (bp) and three exons and two introns in it (Figure 1). The peptide coding regions are separated in two exons by a long intron of 786 bp. Thus the two coding regions of insulin (B chain and junction with C-peptide and C-peptide junction with A chain) are widely separated from each other. Intervening sequences (introns) are deleted during maturation of messenger-RNA.

It is well known that transcription of the gene to messenger-RNA is regulated by a certain structure of upstream from 5' end of the gene. The insulin gene might not be the exception. Actually, a report in which in vitro expression of the

FIGURE 1. Human insulin gene. Peptide coding regions are marked; Pre-P: prepropeptide, B: B-chain, C: C-peptide, A: A-chain. IVS indicates introns.

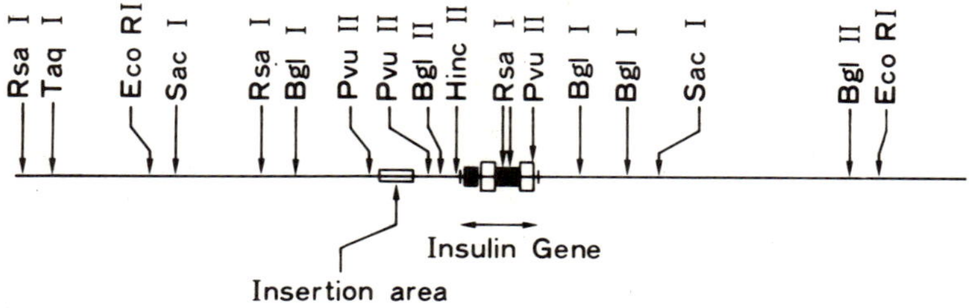

FIGURE 2. Map of human insulin gene adjacent area. Arrows indicate restriction endonuclease cleaving sites.

human insulin gene was observed indicated that the rate of transcription was regulated by the DNA sequence of around 200 bp upstream from 5' end of the insulin gene (6). A restriction endonuclease map is illustrated in figure 2. Analysis of Caucasian population insulin genes using restriction endonuclease revealed that there was polymorphic area between Bgl I and Bgl II cleaving sites flanking the 5' end of insulin gene (7). Detailed analysis of this area showed that this area composed with simple tandemly repeating sequence of 14 DNA bases (ACAGGGGTGTGGGG) and the number of repetitions produced the length polymorphism (8). The meaning of these structures will be discussed in the following section.

III. MUTANT INSULIN GENES

A mutant insulin gene is generally defined as the mutation (in majority cases one point mutation) of an insulin gene. Since mutation of the intron area does not reflect any amino acid sequence of (pro)insulin molecule, none of these mutations can be detected clinically. In addition, alteration of the amino acid sequence of insulin which does not produce any changes in biological activity also cannot be detected clinically. Modification of biological activity of insulin molecule is caused either by amino acid substitution of insulin binding sites (B24-25, A1, and A21) or by the substitution of structurally important amino acids as described in figure 3.

Other important sites for the (pro)insulin molecule are junctions which connect C-peptide and insulin. As described in figure 1, the insulin molecule is synthesized as single chain polypeptide: preproinsulin. The prepeptide part of preproinsulin is removed, soon after the synthesis, in the rough surface endoplasmic reticulum sac. A tertiary structure of (pro)insulin is spontaneously formed by the aid of C-peptide (8). The full biological activity is occasioned by the conversion of proinsulin to insulin catalyzed by specific converting enzyme(s). Important sites of the proinsulin molecule for conversion are junctional sites; arginine(31)-arginine(32) and lysine(64)-arginine(65). The modification of junctional amino acids results in incomplete conversion of proinsulin to insulin and thus accumulates conversion intermediates which attach C-peptide to either side of the insulin moleculer in the pancreatic B-cells and plasma of affected person. These products are also less biologically active as compared to authentic insulin.

These mutant insulins are classified into two clinical forms: 1) increase of circulating abnormal insulin (5800 dalton), or 2) increase of circulating proinsulin like material

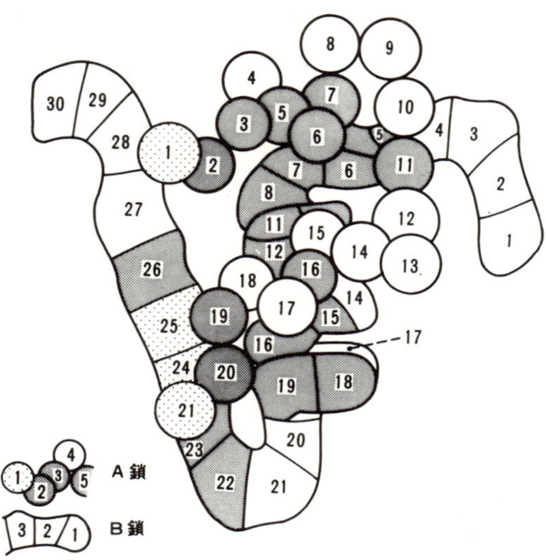

FIGURE 3. Tertiary structure of insulin molecule. Shadowed amino acids are imporant to maintain tertiary structure and pale amino acids are crucial for insulin binding.

(9000 dalton). The former is clinically described as abnormal insulinemia, and the latter as hyperproinsulinemia.

Abnormal insulinemias have been so far reported in 4 cases (families) as summarized in table I. The site of mutation in three of them was identified by cloning of their insulin genes and DNA analysis. All of these patients are heterozygous and inherited autosomal dominant traits. An equal amount of normal and abnormal insulin molecules is expected to synthesize in these patients, because they are heterozygous and have normal and abnormal insulin genes, however more than 90% of immunoreactive insulin (IRI) in peripheral circulation is abnor-

TABLE I. Abnormal insulinemia, reported cases.

Name of Insulin	Abnormality	Mutation on DNA	Ref.
Chicago	B25 Phe - Leu	C - G	(9)
Los Angeles	B24 Phe - Ser	T - C	(10)
Wakayama	A3 Val - Leu	G - T	(11)
Tochigi*	A3 Val - Leu	?	(12)

* Abnormality of the insulin molecule has been determined by amino acid analysis.

mal material. The major part of this production-pripheral circulation discrepancy of percentage of abnormal insulin can be explained by the difference in the rate of degradation of these two molecules.

Insulin Chicago was analysed biochemically at first (13) and the mutation site was determined by analysis of the patient's insulin genes. A point mutation of DNA was found at the site coding B25 amino acid. The change of DNA base from cytidine to guanine indicated the substitution of phenylalanine by leucine(9). Insulin Los Angeles was first analysed by high performance liquid chromatography and found to be different from that of insulin Chicago (14). The abnormality was finally determined by analysis of the patient's insulin genes. Point mutation of thymine to cytidine at the B24 amino acid established change of amino acid from phenylalanine of this site to serine(10). Insulin Wakayama was found in two families, one living in Wakayama and the other in Osaka. Analysis of their insulin genes revealed one point mutation of guanine to thymine at the amino acid A3. The mutation indicates the substitution of amino acid at this site from valine to leucine (11). The fourth case Insulin Tochigi found in Tochigi Prefecture, showed an identical amino-acid substitution observed in Insulin Wakayama. Further studies are needed to elucidate whether this family is related to the families of Insulin Wakayama by common origin.

The metabolic abnormality in the patients having mutant insulin was analysed in two families(10, 11). The age-dependent deterioration of glucose metabolism was observed in these familes (Figure 4).

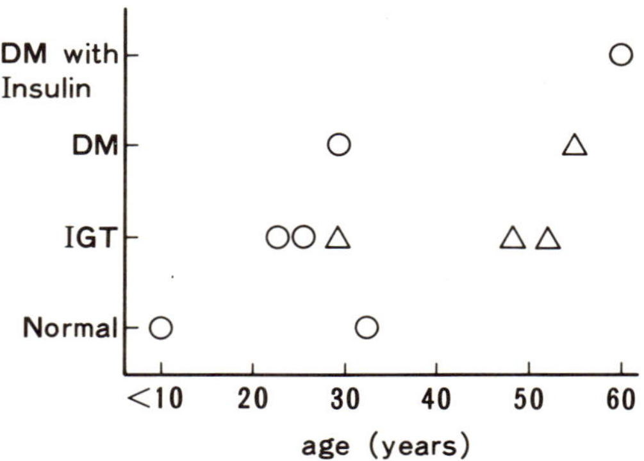

FIGURE 4. The relationship of deterioration of glucose metabolism to the age having mutant insulins. Circle: Insulin Los Angeles and triangle: Insulin Wakayama.

Since these patients have only 50% genetic loss of normal insulin synthesis, it is difficult to assume diabetic metabolism due to the deficiency of 50% lower biological activity of insulin. There should be a further loss of insulin secretion with aging, or an increase of peripheral resistance due to the advancement of age. In this view, these cases will give us useful information for understanding the pathogenic mechanism of non-insulin dependent or adult-onset diabetes.

More cases will be found by detailed analysis of insulin genes in a much bigger population.

Hyperproinsulinemia, on the other hand, was found as an abnormality of elevated proinsulin like material(s) (PLM) in the patients' plasma. Five cases (families) with such abnormality have been described already (Table II). Common clinical features of the Boston (20) and two Tokyo families are: more than 90% of plasma immuno-reactive insulin belonging to PLM; and most of them, except in the case of Toronto whose type of inheritance is not yet identified, are inherited in autosomal dominant traits, whereas New Jersey cases had 70% of PLM in their plasma IRI (21). The Toronto case was much more complicated. The patient, an old lady, had more than 90% of PLM in her plasma at the begining of the study. However, percentage of PLM in her plasma changed markedly after the treatment of glucocorticoid due to the complication of idiopathic thrombopenic purpura. The propositus of the Toronto, Boston and New Jersey families was detected because of the presence of hypoglycemia and they were therefore thought to have insulinoma. The two Tokyo cases were found due to the mild diabetes with hyperinsulinemia by immunoassay of the patients' plasma.

As described in Table II, mutation sites in two cases were determined by the analysis of their insulin genes.

The New Jersey case is somewhat particular as hyperproinsulinemia because it does not have any abnormality in junctions but in the B-chain of insulin. This type of abnormality can be called abnormal insulinemia. However, B10 histidine

TABLE II. Hyperproinsulinemia, reported cases.

Place where family found	Affected junction	Abnoramlity (Mutation in Gene)		Ref.
Toronto	(C - A) ?	?		(15)
Boston	(B - C) ?	?		(16)
Tokyo	(C - A)	Arg - His	(C - A)	(17)
New Jersey	none*	His - Asp	(C - G)	(18)
Tokyo-II	?	?		(19)

* Mutation existed in B10 amino acid.

existed at the important turning point of the alpha-helix of the B-chain, thus substitution of amino-acid in this place might modify the tertiary position of two junctions as well as the structure of insulin. The inhibition of the conversion occurred in this manner.

Our case is rather straight forward one. Propositus, an old man, was found because he had mild diabetes and hyperinsulinemia. Fractionation of his plasma revealed more than 90% of PLM in plasma IRI (22). Chemical analysis of plasma PLM indicated the abnoramlity existed at arginine 65 (23). Analysis of patient's insulin genes confirmed the biochemical results and cytidine to adenine mutation was proved. The substitution of arginine at this site by histidine inhibited the action of converting enzyme at this site (17). Clinical studies of our family revealed again age-dependent deterioration of glucose tolerance in patients who have hyperproinsulinemia. This result again suggests that hyperproinsulinemia might be a good model to study the pathogenesis of adult-onset non-insulin dependent diabetes mellitus. These cases of hyperproinsulinemia might also give us important information for the analysis of converting enzyme(s) which is not clarified yet.

IV. POLYMORPHISM OF INSULIN GENE AREA

Two types of polymorphisms have been observed around the insulin gene. One is restriction fragment length polymorphism, and another is the sites polymorphism which is revealed by the treatment of certain endonuclease (for example Taq I or Rsa I) (24).

Initial observation suggested the significant relationship between long-insertion (more than 120 repetition of tandem sequences) and non-insulin-dependent diabetes (NIDDM) (25). However, later closer relationship of short insertion (less than 40 repetition) with insulin-dependent diabetes mellitus (IDDM) was reported (26). Although there have been many reports concerning the metabolic condition of patients who have had either length of insertion, almost all of them pose difficulties in finding a certain significant connection of these length polymorphisms to insulin response, HbAlc, and other metabolic indices. The only one which has a weak connection was hyperlipidemia (27).

We have studied restriction fragment length polymorphism in Japanese population and both types of diabetics. The results were quite different from those of Caucasian cases (28). We have a much smaller number of long insertion cases than those observed in either Caucasians (25) or American Blacks (26). Our results are compared with those observed in Caucasi-

TABLE III. The restriction fragment length polymorphism of insulin gene in Caucasians (25) and Japanese (28).

Group	Caucasian		Japanese	
	Class 1*	Class 3	Class 1	Class 3
Nondiabetics	0.73	0.26	0.98	0.02
IDDM	0.84	0.16	1.00	0.00
NIDDM	0.66	0.34	0.96	0.04

* Class 1 is short and class 3 is the long insertion.

ans and described in Table III. This marked difference of frequencies of each allele is interesting in view of the genetic evolution of eukariotic genes. However, there is no common relationship of this polymorphism and diabetes in both races. This polymorphism is not a useful marker of diabetes of either type of diabetes in Japanese.

Concerning the sites polymorphism, we do not have enough data to discuss the relationship with diabetes for the moment.

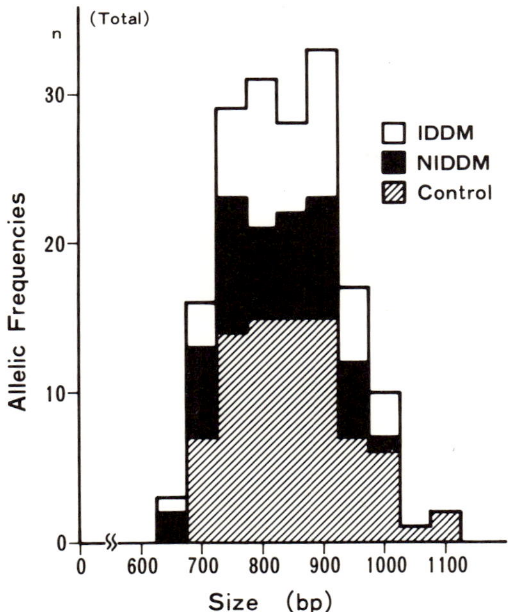

FIGURE 5. Restriction fragment length polymorphism using Pvu II in Japanese.

Further studies were carried out about the short insertion of insulin genes in the Japanese population. The data obtained from unrelated individuals of non-diabetics, IDDM and NIDDM are shown in Figure 5. There are certain polymorphisms which can be detected with Pvu II in these three groups. However the distribution of these three groups is almost the same and we could not detect any significant difference between them. Pvu II fragment of 5' end flanking area of the insulin gene could be a useful marker of the insulin gene to follow in a family tree.

Considering the heterogeneity of the diabetic syndrome, we still have some hope to be able to detect some relationship between insulin gene area polymorphism and certain type of diabetes. For this purpose family study will give us more useful information about the classification and cause-disease relationship of diabetes.

REFERENCES

1. Neel, J. V., in "The Genetics of Diabetes Mellitus" (W. Creutzfeldt, J. Koebberling, and J. V. Neel eds), P 1. Springer Verlag, Berlin 1976
2. Ebina, Y., in "New Approach to Genetic Disease" (T. Sasazuki ed.), 63-68pp, Academic Press Japan Inc. Tokyo 1987
3. Owerbach, D., Bell, G. I., Rutter, W. J., Brown, J. A., and Shows, T. B., Diabetes 30:267 (1981)
4. Bell, G. I., Pictet, R. L., Rutter, W. J., Cordell, B., Tischer, E., and Goodman, H. M., Nature 284:26. (1980)
5. Walker, M. D., Edlund, T., Boulet, A. M., and Rutter, W. J., Nature 306:557. (1983)
6. Bell, G. I., Karam, J. H., and Rutter, W. J., Proc. Natl. Acd. Sci. USA, 78:5759 (1981)
7. Bell, G. I., Selby, M. J., and Rutter, W. J., Nature 295:31. (1982)
8. Kwok, S. C. M., Steiner, D. F., Rubenstein, A. H., Tager, H. S., Diabetes 32:872. (1983)
9. Haneda, M., Chan, S. J., Kwok, S. C. M., Rubenstein, A. H., and Steiner, D. F., Proc. Natl. Acad. Sci. USA, 80: 6366. (1983)
10. Nanjo, K., Sanke, T., Miyano, M., Okai, K., Sowa, R., et al., J. Clin. Invest. 77:514. (1986)
11. Iwamoto, Y., Sakura, H., Yui, R., Fujita, T., Sakamoto, Y., Matsuda, A., and Kuzuya, T., Diabetes, 35:suppl. 1, 77A #306. (1986)

12. Given, B. D., Mako, M. E., Tager, H. S., Baldwin, D., Markese, J., Rubenstein, A. H., Olefsky, J., Kobayashi, M., Kolterman, O., and Poucher, R., N. Engl. J. Med. 302: 129. (1980)
13. Shoelson, S., Haneda, M., Blix, P., Nanjo, K., Sanke, T., Inouye, K., Steiner, D. F., Rubenstein, A. H., and Tager, H. S., Nature, 302:540. (1983)
14. From, G. l. A., and Kahn S., Diabetes 20:suppl. l. 335. #37 (1971)
15. Gabbay, K. H., Bergenstal, R. M., Wolff, J., Mako, M. E., and Rubenstein, A. H., Proc. Natl. Acad. Sci. USA 76:2881 (1979)
16. Shibasaki, Y., Kawakami, T., Kanazawa, Y., Akanuma, Y., and Takaku, F., J. Clin. Invest. 76:378. (1985)
17. Chan, S. J., Seino, S., McCormick, M. B., Gruppuso, P. A., Schwartz, R., Steiner, D. F., Diabetes 35:suppl 1. 97A. #387, (1986)
18. Ohashi, H., Kanemitsu, M., Omori, M., Ohgawara, H., Odagiri, R., Mihara, T., Hirata, Y., Haneda, M., Folia Endocri. Jap. 60:suppl, 1078#181 (1984)
19. Gabbay, K. H., DeLuca, K., Fisher, J. N. Jr, Mako, M. E., and Rubenstein, A. H., N. Engl. J. Med. 294:911. (1976)
20. Gruppuso, P. A., Gorden, P., Kahn, R., Cornblath, M., Zeller, P., and Schwartz, R., N. Engl. J. Med. 311:629 (1984)
21. Kanazawa, Y., Hayashi, M., Ikeuchi, M., Kasuga, M., Oka, Y., Sato, H., Hiramatsu, K., and Kosaka, K., in "Proinsulin, Insulin, C-peptide" (S. Baba, T. Kaneko, and N. Yanaihara eds.) P 262. Excerpta Medica Amsterdam ICS 468 1979
22. Robbins, D C., Blix, P. M., Rubenstein, A. H., Kanazawa, Y., Kosaka, K., and Tager H. S., Nature, 291:679. (1981)
23. Elbein, S. C., Corsetti, L., and Permutt, M. A., Diabetes, 34:1139. (1985)
24. Rotwein, P. S., Chirgwin, J., Province, M., Knowler, W. C., Pettitt, D. J., Cordell, B., Goodman, H. M., and Permutt, M. A., N. Engl. J. Med. 308:65. (1983)
25. Bell, G. I., Horita, S., and Karam, J. H., Diabetes, 33: 176. (1984)
26. Jowett, N. I., Williams, L. G., Hitman, G. A., and Galton, D. J., Brit. Med. J. 288:96. (1984)
27. Awata, T., Shibasaki, Y., Hirai, H., Okabe, T., Kanazawa, Y., and Takaku, F., Diabetologia, 28:911. (1985)

CLONING AND EXPRESSION OF A cDNA CODING FOR THE HUMAN INSULIN RECEPTOR

Yousuke Ebina

Institute for Medical Genetics
Kumamoto University Medical School
Kumamoto

I. INTRODUCTION

Diabetes mellitus is caused either by a deficiency of insulin or by insensitivity of the target cell to insulin. This latter defect, which is mainly inherited, causes the majority of diabetes. An understanding of the defect in these patients requires an understanding of the molecular mechanisms of insulin action. Since the manifold insulin responses in target cells are initiated by the binding of insulin to its receptor, we cloned the human insulin receptor cDNA.

II. RESULTS AND DISCUSSION

A. Cloning the Human Insulin Receptor cDNA

We purified the human insulin receptor protein. The receptor is composed of two subunits, α and β whose molecular weight are about 135K and 90K. Both subunits are highly glycosylated. To increase the chance of isolating the 5' portion of a presumably long (>5kb) mRNA, we constructed two cDNA libraris: a random primed cDNA library, and an olig(dT)-primed cDNA library using the λgt11 vector system (1). Based on the N-terminal amino acid sequence information of the α subunit, we obtained the entire cDNA of the receptor from these cDNA libraries. Then, we determined the complete nucleotide sequence of the human insulin receptor cDNA (2).

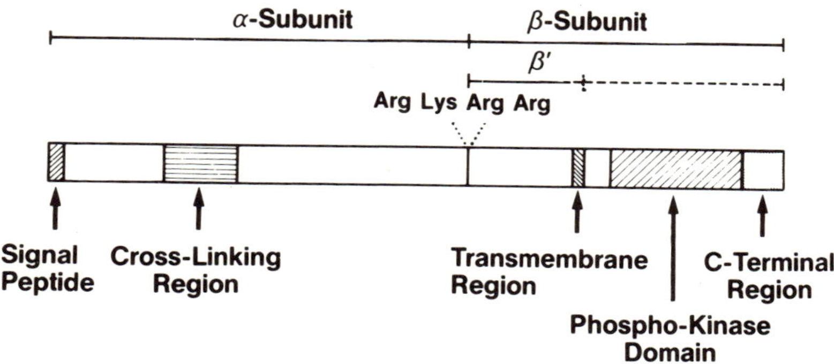

FIGURE 1. Structure of the human insulin receptor

The structure of the receptor is shown based on the predicted amino acid sequence of the cDNA (Fig.1). The single precursor of the receptor is synthesized. The precursor is proteolytically cleaved into α and β subunits after the segment Arg-Lys-Arg-Arg whiich is accessible to processing enzyme. In the α subunit, there are signal peptide, and cystein rich, cross-linking region. The insulin binding site is somewhere in the α subunit. In the β subunit, there are transmembrane region and tyrosine kinase domain. The region left from the transmembrane region is extracellular, and the rest of the receptor is inside the cell. The α and β subunits are linked by S-S bond.

The chromosomal location of the gene was analyzed by direct hybridization to the DNA of chromosomes which are resolved by dual laser chromosome sorting (3). The human insulin receptor gene mapped to chromosome 19. Interestingly, this chromosomal location is shared by another membrane receptor, the LDL receptor.

In general, the insulin receptor shares many structural and functional features with the EGF receptor (4). Sequence comparison of the insulin receptor β subunit and the corresponding domain of EGF receptor to different viral transforming proteins reveals homology between these receptors and oncogenic proteins. Of all the known oncogenic proteins, the v-ros gene product (5) is the most homologous to the insulin receptor. To elucidate the mechanisms of insulin action, the elements of the insulin receptor must somehow be functionally isolated and reconstituted. Then we expressed the functional human insulin receptor in a heterologous cell system (6).

B. Expression of the human insulin receptor cDNA

The human insulin receptor expression plasmid was cotransfected with a plasmid conferring resistance to the neomycin. The neomycin-resistant colonies were picked randomly. The clones were assayed for the presence of the human insulin receptor on the cell surface by selective binding of a monoclonal antiboby specific for the human insulin receptor (7). CHO cells have their own insulin receptor. To examined the specific binding and action of insulin through the human insulin receptor in the transfected cells, we used a monoclonal antibody specific for the human insulin receptor. The monoclonal antibody blocks the insulin binding and action (7). Thus the antibody allows us to discriminate between the CHO receptors and the human receptors.

We examined the specificity of ligand binding by assaying binding of ^{125}I-labeled insulin to the transfected cells (CHO-HIR3) and control CHO cells. In the absence of unlabeles insulin, 2% of the ^{125}I-labeled insulin is specifically bound by the transfected cells, whereas only 0.5% was bound by control CHO cells (Table I). The binding of ^{125}I-labeled insulin to both of these cell lines was inhibited in a dose-dependent fashion by unlabeled insulin. The human receptor-specific monoclonal antibody blocked insulin binding to the transfected cells but not CHO cells. It indicates that the increased binding observed in the transfected cells is via the human insulin receptor. In contrast, both cell types bound 0.4% of ^{125}I-labeled IGF-1. A Scathard plot analysis suggests that under these assay conditions the transfected cells display approximately 14,000 and CHO cells about 2800, high-affinity sites per cell. The apparent dissociation constants of both receptors are about 5×10^{-10}M.

TABLE I. ^{125}I-Insulin Binding to CHO-HIR3 and Control CHO Cells

	^{125}I-Insulin Bound (% total)						
	Insulin (M)						
	0	10^{-11}	10^{-10}	10^{-9}	10^{-8}	10^{-7}	10^{-5}
CHO-HIR3	2	1.9	1.7	0.9	0.45	0.3	0.25
CHO	0.6	0.55	0.48	0.3	0.2	0.15	0.1

TABLE II. Autophosphorylation of the Partially Purified Insulin Receptor from CHO-HIR3 and Control CHO cells

	Incorporation of ^{32}P into the β subunit (cpm)	
	-Insulin	+Insulin
CHO-HIR3	5459	12609
CHO	521	1125

We next examined whether the insulin receptor from the transfected cells displays insulin-activated tyrosine kinase activity of the β subunit (Table II). In partially purified total cellular insulin receptor preparations from both the transfected and CHO cells, the molecular weight 95,000 β subunit of the receptor was exclusively phosphorylated. The incorporation of ^{32}P into the β subunit was 10 times greater in the transfected cells than in CHO cells in both the presence and the absence of insulin. Insulin stimulated the observed phosphorylation 2-fold in both the transfected cells and the control CHO cells.

We assayed the insulin stimulation of glucose uptake to assess whether the human insulin receptor expressed in the transfected cells exhibits a physiological response. The maximal stimulation by insulin of glucose uptake by both the transfected and CHO cells was similar, but for one-half the maximal response, CHO cells required at about 30 times greater concentration of insulin than the transfected cells. The increased sensitivity is not due to an increased affinity, since the estimated dissocaiation constants for the human and the hamster insulin receptors are very similar. From the apparent dissociation constant, we calculated that about 17% of the cell surface receptors must be occupied for half-maximal response in CHO cells, whereas only 0.8% of the total receptors must be occupied for half-maximal response in the transfected cells. Thus, there are "Spare Receptors" in both cell types.

TABLE III. Insulin Stimulation of 2-Deoxy[^{3}H] Glucose Uptake in a CHO-HIR3 and Control CHO cells

	2-Deoxy[^{3}H] Glucose Uptake (cpm)				
	Insulin (M)				
	0	10^{-11}	10^{-10}	10^{-9}	10^{-8}
CHO-HIR3	3000	5200	5800	5900	6000
CHO	3000	3400	4600	5900	6100

The number of occupied receptors required for the biological response appears to be significantly different in the two cell types. More effective clusters of the receptors might be generated at the higher receptor concentrations present in the membrane of the transfected cells.

The ability to express the human insulin receptor in this heterologous cell system permits the further analysis of structure-function relationships of the receptor molecule and of the mechanisms of insulin action. And also the cDNA is available for the screening of the insulin receptor gene abnormality in diabetes patients.

ACKNOWLEDGMENTS

This work was done in U.C.S.F. with many collaborators. I especially acknowledge the contributions of L. Ellis, K. Jarnagin, M. Edery, L. Graf, E. Clauser, J. Ou, D. Standring, F. Masiars, Y.W. Kan, I. D. Goldfine, J. Beaudoin, R.A. Roth and W.J. Rutter. I also would like to express my thanks to my present collaborators, especially, Drs. E. Araki, M. Taira, N. Shimada and Prof. M. Mori in Kumamoto University School of Medicine.

REFERENCES

1. Young, R.A., and Davis, R.W. (1983). Science 222:778
2. Ebina, Y., Ellis, L., Jarnagin, K., Edery, M., Graf, L., Clauser, E., Ou, J.-H., Masiarz, F., Kan, Y.W., Goldfine, I. D., Roth, R.A., and Rutter, W.J. (1985). Cell 40:747.
3. Lebo, R., Gorin, F., Fletterick, R.J., Kao, F.T., Cheung, M.C., Bruce, B. I., and Kan, Y.W. (1984). Science 225:57.
4. Ullrich, A., Coussens, L., Hayflick, J. S., Dull, T.J., Gray, A., Tam, A.W., Lee, J., Yarden, Y., Libermann, T.A., Schles singer, J., Downward, J., Mayes, E.L.V., Whittle, N., Waterfield, M.D., and Seeburg, P.H. (1984). Nature 309:418
5. Shibuya, M., Hanafusa, H., and Balduzzi, P.C. (1982). J. Virol. 42:143.
6. Ebina, Y., Edery, M., Ellis, L., Standring. D., Beaudoin, J., Roth A.R., and Rutter, W.J. (1985). Proc. Natl. Acad. Sci. USA 82:8014
7. Roth, R.A., Cassell, D.J., Wong, K.Y., Maddux, B.A., and Goldfine, I.D. (1982). Proc. Natl. Acad. Sci. USA 79:7312

SUPRATYPES AS MARKERS OF DISEASE GENES AND MAJOR HISTOCOMPATIBILITY COMPLEX CLASS III GENES

Subtitle: Comparison of IgA deficiency insulin-dependent diabetes mellitus and myasthenia gravis

Roger L Dawkins, E. Martin,
T.J. Cobain, P.H. Kay and F.T. Christiansen

Publication Q8649 of the Departments of Clinical Immunology, Royal Perth Hospital, The Queen Elizabeth II Medical Centre and the University of Western Australia, Perth, Western Australia.

INTRODUCTION

After more than a decade of cataloguing HLA-disease associations there is a surfeit of information but a disappointing lack of clear and simple explanations. For example, IgA deficiency has been associated with more than 20 alleles at HLA A, B or DR loci but none of these alleles can be regarded as directly relevant to the pathogenesis. Elsewhere (Wilton et al., 1985) we have shown that at least most of these alleles can be explained in terms of only three supratypes as listed in Table 1.

It follows that these supratypes must be the best markers of the relevant disease genes and that these genes could be anywhere within or close to the MHC. Accordingly our strategy has been:

(i) to develop cell lines which are homozygous for one of the relevant supratypes or heterozygous for any two,

and (ii) to use these cell lines as a means of identifying the location and function of relevant genes.

TABLE 1

DISEASE ASSOCIATED SUPRATYPES

SUPRATYPE					DISEASE		
B	C4A	C4B	Bf	DR	IgA Def.	IDDM	MG
7	3	1	S	3	-	-	-
8	QO	1	S	3	+	+	+
18	3	QO	F1	3	-	+	-
65	2	1+2	S	1	+	-	-

With the intention of identifying disease genes, we have compared supratypes associated with IgA deficiency with those associated with other diseases such as insulin dependent diabetes mellitus (IDDM) and myasthenia gravis (MG). In fact A1 Cw7 B8 C4B1 C4AQ0 BfS DR3 DQ2 is common to all three diseases, whereas Bw65 C4B1+2 C4A2 BfS is associated with IgA deficiency but not IDDM or MG. B18 C4BQ0 C4A3 BfF1 DR3 DQ2 is associated with IDDM but not MG or IgA deficiency. A comparison of these different supratypes might reveal elements relevant to one or other of these diseases.

To test the importance of any interesting characteristics it should be possible to take advantage of the fact that these diseases appear to be at least similar in different racial groups even though the HLA associations are different. Assuming that these differences are due to different associations between HLA alleles and the putative disease genes ("linkage disequilibrium") we have chosen to compare the restriction enzyme maps in Caucasian and Mongoloid populations.

IgA deficiency is extremely rare in Mongoloids but we have previously shown that Chinese patients with MG may have HLA B8 even though this antigen is extremely rare in the Chinese population as a whole (Garlepp et al., 1983; Christiansen et al., 1984). We have therefore used the presence of autoimmune disease as a means of ascertaining subjects with B8 and we have asked whether such subjects resemble Caucasians with A1 B8 DR3 in terms of associated genes within the MHC.

TABLE 2

ASSOCIATIONS BETWEEN B8 AND AUTOIMMUNE DISEASES IN CHINESE

Identification	MHC Supratype						Disease
	A	B	C4A	C4B	Bf	DR	
8439	2,32	8,40	Q0,3	1	S	3,9	Graves' disease
7655	2,26	8,13	Q0,3	1	F,S	N/D	IDDM
8152	1,19	8,54	3	1,5	S	3,9	
7025	11	7,8	3	1	F,S	N/D	SLE
8441	11,26	8,40	3	1	F,S	3	SLE
8515	11,26	8,17	3	1	F,S	3	Renal Disease
7889	2,26	8,46	4	1,2	F,S	2,9	MG
7564	1,19	8,15	4	2	S	N/D	MG
7563	11	8,15	3	1,2	F,S	N/D	Graves' disease

MATERIALS AND METHODS

The cell lines were prepared from subjects with relevant supratypes as shown by complete MHC genotyping. Details have been given elsewhere (Grimsley et al., 1986).

The Chinese patients have various autoimmune diseases as listed in Table 2. All had been shown to have B8. In one case an EBV transformed cell line was produced but in the other instances DNA was extracted from peripheral blood cells as previously described (Garlepp et al., 1986).

MHC typing

Methods for HLA and complement allotyping have been described elsewhere (Garlepp et al., 1986).

Restriction enzyme analysis

Methods are essentially as described by Garlepp et al. (1986) with the following modifications. Blots were washed in 2 changes of 2xSSC, 0.5% SDS for 5 minutes at room temperature, followed by two 30 minute washes in 1xSSC, 0.5% SDS at 68°C. pC4AL1 is a 0.95Kb cDNA probe which codes for the C4 gamma chain (Whitehead et al., 1983).

RESULTS

Comparison of Class III regions: In the first instance we undertook a comparison using the 21-hydroxylase probe. Since the A3 Cw7 B7 C4B1 C4A3 BfS DR2 supratype is negatively associated with IDDM (and possibly also MG and IgA deficiency), we have used this supratype as a reference (Figure 1).

Inter-racial comparisons: The Chinese subjects with B8 appear to resemble Caucasians with 1,8,3 in that the Taq I 3.2Kb fragment was reduced relative to the 3.7Kb fragment (Figure 1).

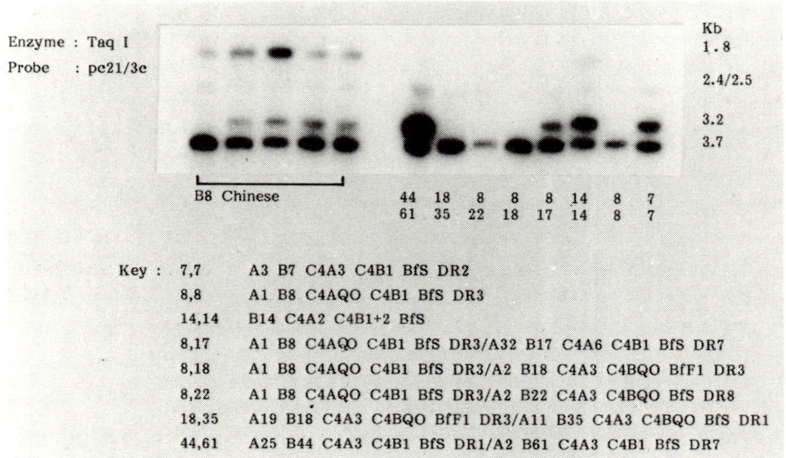

Figure 1. Restriction fragments detected in genomic DNA from Chinese patients with HLA B8 and various Caucasoid cell lines treated with Taq I and hybridized with the 21-OH probe.

The standard on the right of the figure (lane 1) contains the expected 3.7, 3.2Kb fragments together with a doublet at 2.4 and 2.5Kb. By contrast the 1,8,3 supratype lacks the 3.2 and 2.4Kb fragments (lane 2). In the B14 containing supratype there is increased density of the 3.2 and 2.4Kb fragments together with an additional fragment at approximately 1.8Kb (lane 3). In lane 4 the pattern is explained by the presence of the B8 containing supratype. In lane 5 heterozygosity for the B8 and B18 containing supratypes explains the absence of the 3.2 and 2.4Kb fragments. In lane 6 the pattern is very similar, suggesting that the second haplotype (containing C4BQ0 and C4A3) also contains a deletion of 21-A. In lane 7 apparent deletions of 21-A are also associated with C4BQ0 and C4A3. Lane 8 contains more DNA but has been shown elsewhere to have a heterozygous deletion of the 3.7Kb fragment (Dawkins et al., unpublished).

In lanes 10-14 inclusive it can be seen that the Chinese cells have a reduction in the 3.2Kb fragment. In addition it can be seen that there is a band at approximately 1.8Kb. Lane

14 contains DNA which appears to be homozygous for a deletion of 21-A.

Furthermore, there appeared to be a single fragment at 2.4 or 2.5Kb in contrast to the doublet seen in the Caucasian standard. In these two respects it would appear that the Chinese with B8 may lack the 21A gene. On the other hand, the Chinese patients had an extra fragment at 1.8Kb. This fragment is not found in the Caucasian standard; nor is it found with 1,8,3 or 18,F1,3. On the other hand it has been seen in a cell line which is homozygous for Bw65 C4B1+2 C4A2 BfS (lane 3), suggesting a possible similarity between Chinese with B8 and the Caucasian supratype associated with IgA deficiency.

Extent of the similarity between Chinese and Caucasian subjects with B8: The phenotypes of the Chinese subjects are shown in Figure 2. It can be seen that the Chinese patients resemble the Caucasians with respect to the class III region and possibly also in the HLA-D region. By contrast the Chinese patients generally differed in that they did not have HLA A1.

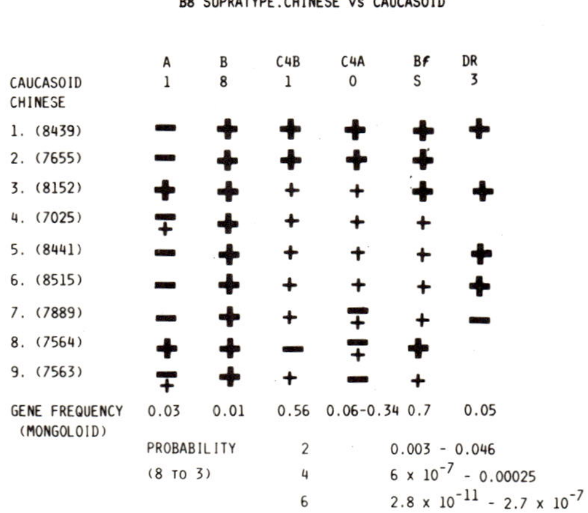

B8 SUPRATYPE.CHINESE vs CAUCASOID

	A	B	C4B	C4A	Bf	DR
CAUCASOID	1	8	1	0	S	3
CHINESE						
1. (8439)	−	+	+	+	+	+
2. (7655)	−	+	+	+	+	
3. (8152)	+	+	+	+	+	+
4. (7025)	∓	+	+	+	+	
5. (8441)	−	+	+	+	+	+
6. (8515)	−	+	+	+	+	+
7. (7889)	−	+	+	∓	+	−
8. (7564)	+	+	−	∓	+	
9. (7563)	∓	+	+	−	+	
GENE FREQUENCY (MONGOLOID)	0.03	0.01	0.56	0.06-0.34	0.7	0.05

PROBABILITY (8 TO 3)		
	2	0.003 - 0.046
	4	6×10^{-7} - 0.00025
	6	2.8×10^{-11} - 2.7×10^{-7}

Figure 2. Phenotypes of Chinese patients with B8.

It can be seen that at least most of these patients are very similar to Caucasians in that they have C4B1 C4AQ0 BfS and DR3. By contrast only 2 of the 9 have A1. These results suggest that B8 in Chinese patients marks a segment of DNA and that the relevant segment extends through the Class III region and possibly into the D region. Apparently this segment represents racial admixture. It is concluded that the relevant genes could be anywhere within the B-D segment, possibly close to the Class III region.

DISCUSSION

These results extend our analysis of supratypes associated with disease. Our approach is based upon the assumption that combinations of alleles can be used to mark extensive segments of the MHC and that these segments may be more important than the individual alleles recognised by conventional HLA typing. For example, there may be cis interactions between the different genes as might occur if the D region controlled antigen specific responses and the Class III region controlled aspects related to complement and hormonal responsiveness. Furthermore, there may be as yet unidentified genes within these segments.

The 1,8,3 supratype is common to IgA deficiency, IDDM and MG. As reported by others (Carroll et al., 1985; White et al., 1985) and as confirmed previously (Garlepp et al., 1986) there is an extensive deletion involving the 21-A gene. The present results assist in characterising the extent of this deletion. By defining the restriction map for a standard with 3,7,2 we have been able to clarify some previous interpretations which have suffered from the lack of a homozygous reference cell. There are several possibilities but a simple explanation for the RFLPs described here would be a deletion from within C4A to within C4B. Such a deletion implies the presence of a recombinant gene which can be regarded as a hybrid of C4A and C4B. Alternatively the C4 gene associated with 1,8,3 might be regarded as primordial implying a reduplication event to explain the presence of C4A and C4B and 21-A and 21-B in the standard 3,7,2 supratype. In either case our findings raise the possibility of unequal crossing-over with combining sites within rather than between the relevant genes. Given the extensive homology between C4A and C4B and between 21-A and 21-B it would appear quite probable that deletions and duplications will occur. However, it should be emphasised that these recombinants are relatively stable since the appropriate RFLP patterns can be predicted from the phenotype in most, if not all, cases. It follows that there may be some important functional implications of these arrangements at the DNA level.

The possibility of hybrid genes wihin the Class III region suggests a mechanism for the generation of extensive polymorphism. For example, there may be multiple hybrid products in addition to the polymorphism reflected by the currently recognised alleles at C4A and C4B.

The 18,F1,3 supratype is similar to the 1,8,3 supratype in containing a deletion of 21-A but differs in two major respects. Firstly, 1,8,3 is typed as C4B1 and C4A null whereas 18,F1,3 is typed as C4A3 and C4B null. Secondly, restriction enzyme analysis reveals that although there are similarities there are differences which suggest that the two supratypes could share the beta region of C4A and the gamma region of C4B. It is therefore possible that the deletions are of similar size but differ in their combining sites. Further, although the C4 product of 18,F1,3 is typed as C4A3, there is evidence that it may actually be different from the C4A3 produced by 3,7,2. By digesting with Bam HI and probing with a C4 gamma probe, different fragments are seen (Uko et al., unpublished). Indeed, it might be predicted that many of the products which occur in association with an apparent null at an adjacent C4 locus will differ from each other in that they will be different hybrids. These differences would be expected to be predictable from supratypes. Whether these differences occur within coding regions and affect the product and the function of the product remains to be determined, but there is at least the potential for differences which could influence complement function and hence susceptibility to autoimmune and other diseases.

The 1,8,3 and the 18,F1,3 supratypes are associated with IDDM and must therefore be similar in some respects. They differ at the Class I loci so far examined and it is now clear that they differ in the Class II region even though each expresses DR3 serologically. As shown by Honeyman et al. (Personal communication) the MLC reactivity is different and there is evidence suggesting differences by restriction enzyme analysis. The two supratypes differ at the Bf locus so that the only similarities now apparent would appear to be in the C4A beta and C4B gamma sub-regions and 21 regions of the MHC. It remains to be determined how similar the two supratypes are at the C2 locus but perhaps the most striking similarity is that each supratype has an extensive deletion encompassing 21-A. Whether the similarities are relevant to IDDM is unknown but it should be noted that there are at least quantitative differences between 1,8,3 and 18,F1,3. As we have shown previously (Kelly et al., 1985), there is evidence that the genes marked by these two supratypes interact differently, possibly reflecting different cis interactions.

The 1,8,3 and 18,F1,3 supratypes differ in that only the former is associated with IgA deficiency and MG. We therefore predict that the genes involved in these disorders are related to the differences rather than the similarities. One of the differences between these supratypes is the presence of the C4B alpha region in the 1,8,3 supratype compared to the C4A alpha region in the 18,F1,3 supratype. The loss of the C4A alpha region in 1,8,3 supratype could be relevant to the association with IgA deficiency and MG.

In a further attempt to localise the relevant genes on the 1,8,3 supratype we have compared the pattern seen in Caucasians with that seen in Chinese patients with B8 and autoimmune diseases. Several years ago we showed that Thai patients with MG could have HLA B8 although this allele is extremely rare in the Thai and Chinese populations (Christiansen et al., 1984). We therefore predicted that there had been some racial mixture and that the relevant MHC genes were marked by B8 and presumably rather close to the HLA B locus. The present data suggest that the relevant gene is in the B-D region and possibly close to the Class III region. At least most of the Chinese patients with B8 resemble Caucasians in their complotypes and in their restriction enzyme pattern seen with Taq I and the 21-hydroxylase probe. Several possibilities need to be considered. Firstly the results are in keeping with the possibility that complement deficiency _per se_ may be relevant, at least in the case of SLE and possibly other autoimmune diseases (Dawkins et al., 1983; Christiansen et al., 1983; Fielder et al., 1983). Secondly, there may be one or more genes within the B-D region yet to be identified. Thirdly, _cis_ interaction may be important and may explain why the rather extensive region marked by HLA B and the complement alleles has been retained in Chinese patients who develop autoimmune diseases.

ACKNOWLEDGEMENTS

We are grateful to Dr B.R. Hawkins for samples from Chinese patients with B8, Drs M.J. Garlepp and L. Taylor for helpful discussion and E. Ly for technical assistance. A preliminary report has been published elsewhere (Cobain, T.J., Kay, P.H., Taylor, L., Dawkins, R.L., Hawkins, B.R., Proceedings of 3rd AOHWC, 1986).

REFERENCES

Belt, K.T., Carroll, M.C., and Porter, R.R. (1984). Cell _36_:901.

Carroll, M.C., Palsdottir, A., Belt, K.T., Porter, R.R. (1985). EMBO Journal. _4_, 10:2547.

Christiansen, F.T., Dawkins, R.L., Uko, G., McCluskey, J., Zilko, P.J. (1983). Aust. NZ J.Med. _13_:438.

Christiansen, F.T., Pollack, M.S., Garlepp, M.J., Dawkins, R.L. (1984). J. Neuroimmunology 7:121.

Dawkins, R.L., Christiansen, F.T., Kay, P.H., Garlepp, M.J., McCluskey, J., Hollingsworth, P.N., Zilko, P.J. (1983). Immunol. Rev. 70: 5.

Fielder, A.H.L., Walport, M.J., Batchelor, J.R., Rynes, R.I., Black, C.M., Dodi, I.A., and Hughes, G.R.V. (1983). Brit. Med. J. 286:425.

Garlepp, M.J., Christiansen, F.T., Dawkins, R.L., Chiewsilp, P. and Takata, H. (1982). In "Immunogenetics in Rheumatology" (R.L. Dawkins et al, eds.) pp 259-263. Excerpta Medica, Amsterdam.

Garlepp, M.J., Wilton, A.N., Dawkins, R.L., White, P.C. (1985). Immunogenetics 23:100.

Grimsley, G., Homes, P., Townend, D., Dawkins, R.L., Christiansen, F.T. (1986). Proc. 3rd AOHW Conference (in press).

Kelly, H., McCann, V.J., Kay, P.H., Dawkins, R.L. (1985). Immunogenetics, 22:643.

White, P.C., Grossberger, D., Onufer, B.J., Chaplin, D.D., New, M.I., Dupont, B., Strominger, J.L. (1985). Proc. Natl. Acad. Sci. USA, 82:1089.

Whitehead, A.S., Goldberger, G., Woods, D.E., Markham, A.F., and Colten, H.R. (1983). Proc. Natl. Acad. Sci. USA 80:5387.

Wilton, A.N., Cobain, T.J., Dawkins, R.L. (1985). Immunogenetics, 21:333.

GENETICS OF HLA-LINKED DISEASE SUSCEPTIBILITY

Ekkehard D. Albert

Immunogenetics Laboratory
University of Munich
Munich, West-Germany

I. INTRODUCTION

Multiple diseases are associated with the HLA region which spans as little as 2-3% recombination frequency which represents less than 1/1000 of the human genome. Among the HLA associated diseases are as diverse as the following: AS, M. Behcet, Thyroiditis de Quervain, Acute Lymphocytic Leucemia, Idiopathic Hemochromatosis (IH), Psoriasis, Juvenile Diabetes Mellitus, Coeliac Disease, Graves Disease, Myasthenia gravis, Autoaggressive Hepatitis, Rheumatoid Arthritis, Juvenile Rheumatoid Arthritis, Nephrotic Syndrome, Multiple Sclerosis, Narcolepsy, Lupus Erythematosus, Sclerodermia, IgA Deficiency, C21-OHase Deficiency.

The fact that so many disease susceptibility genes are mapped in this very limited space of the genome indicates that there could be a common principle involved in the pathogenesis of many of these disorders.

Except for the two recessive diseases (C21-OHase Deficiency and IH) the mode of inheritance is unclear and the penetrance is relatively low as evidenced for example by the observation of discordant pairs of monozygotic twins, which shows that environmental factors are necessary as a trigger of the disease process. For this reason the HLA-linked disease genes are generally viewed as <u>susceptibility</u> genes. For most of the above mentioned disorders one can assume that the immune system is significantly involved in the pathogenesis. Since the gene products of the HLA region are in various ways playing a significant role in the function of the immune system, it is not far fetched to assume that at least some of the disease susceptibility genes could be <u>identical</u> with HLA genes. Thus completely "normal" HLA genes can under certain (rare) circumstances lead to a pathological immune response. (The price which a highly polymorphic system has to pay for the survival of the species). Alternatively the susceptibility genes could be closely linked to HLA with similar but aberrant immune functions.

II. ASSOCIATION WITH HLA-HAPLOTYPES

The haplotype is defined as the unit of inheritance of the HLA-system. The haplotype stands for the combination of antigens coded for by the different closely linked HLA-loci which is passed on from one generation to the next for example, the haplotype HLA-A1, Cw7, B8, C4AQ0, C4B1, BfS, DR3, DRw52, DQw2 is the combination in "cis" of antigens coded for on the same chromosome. Due to the very close linkage of the HLA-loci, and due to selective forces which have favoured certain combinations over others, a large number of combinations is found in much greater frequency as one would expect on the basis of the single gene frequencies of the alleles involved. This phenomenon of the so called "Linkage disequilibrium" complicates considerably the analysis of HLA-disease associations. If one particular disease is associated with HLA-DR3, and if HLA-DR3 is already in the normal population very closely associated with A1 and B8, then one must expect that also in the disease population one will find a very high frequency of the haplotype A1, B8, DR3, which may be significantly different from the frequency of the haplotype in a normal control population. This however does not necessarily mean that disease susceptibility is coded for by the presence of all three markers of this haplotype, namely A1 and B8 and DR3. In such cases it is very difficult to distinguish between the three principal possibilities:

1. The disease is associated with the combination of more than one marker on this haplotype. This represents in the simple most condition an interaction of at least three points, namely two markers (p. e. B8 and DR3) and the disease. This is a third order interaction.
   In order to prove a third order interaction, it is necessary to have sample sizes which by far exceed those that we are generally accustomed to finding in HLA-disease studies.
2. The disease susceptibility gene is located between two marker loci like between HLA-B and HLA-DR and therefore approximately equally associated both with B8 and with DR3. This possibility would be proven only by the demonstration, that in the disease the haplotypes carrying B8

but not DR3 are increased over those in the control population as well as the haplotype carrying DR3 but not B8. Again such calculations require very large sample-sizes of patients which have been genotyped.

3. The disease susceptibility gene is identical with one of the markers on the haplotype, like per example HLA-DR3. In this case one can demonstrate that haplotypes carrying DR3 but not B8 and A1 are increased in frequency while haplotypes carrying B8 and A1 without DR3 are not increased in frequency.

If one follows the logical principles of William Occam, one would try to explain an observation with the least complicated hypothesis possible. It is more likely that a single determinant has primary associations and the others are secondary. If a disease like Juvenile Diabetes is associated with the haplotype A1, B8, DR3 it is likely that the association with A1 and B8 are due to linkage disequilibrium as one can see that in certain populations the same disease is associated with DR3 but not with B8 or A1. Even if it could be established that in the patients the proportion of the A1, B8, DR3 positive among all the DR3 is significantly greater than in the healthy control population, this would not necessarily mean that A1 and/or B8 are contributing to disease susceptibility. Such a finding could still be explained by the fact that it is a particular variant of DR3 which is responsible for disease susceptibility and which is very frequently found on A1, B8 haplotypes.

Such situations are found in classical HLA serology (the B17 splits, Bw57 on the A1, B57 haplotype and B58 on A33, Bw58) and in RFLP-studies, where the DQw1 subtype DQR2.6 is found in strong association with DR2, B7 and not with other DR2 haplotypes.

III. ASSOCIATION WITH MORE THAN ONE ALLELE OF THE SAME LOCUS

It has been observed that for an increasing number of diseases associations with more than one allele of the same locus can be established. This has long been known for Juvenile Diabetes, for Coeliac Disease, Juvenile Chronic Arthritis and Rheumatoid Arthritis. Such multiple associations may find various explanations which are not mutually exclusive.

A. Disease Heterogeneity

It is possible that within one syndrome there are two different disease entities included, one of which is associated with one and the other with another allele. This maybe the case in Juvenile Diabetes, where DR3 homozygotes might represent a form of the disease which is different from Juvenile Diabetes associated in a DR3,DR4 heterozygote.

B. The Association of Two Different Susceptibility Genes with Two Different HLA-Alleles

This could be true also for Juvenile Diabetes as evidenced by the large excess of DR3 for heterozygotes over the sum of the DR3 homozygotes and DR4 homozygotes.

C. Association of the Same disease Susceptibility Gene with Two Different HLA-Genes

This is a possibility for which so far no cogent example has been found.

IV. HETEROZYGOTE EFFECT

The fact that in Diabetes and CD the heterozygote types DR3/4 and DR3/7 respectively have been found to carry the greatest relative risk had led to considerable discussion. A simple calculation shows that there must be two different disease susceptibility genes associated with DR3 and DR4: If the same disease susceptibility gene would be associated with DR3 and DR4 one would expect to find DR3 homozygote and DR4 homozygote = DR3/4 heterozygotes. As can be seen in Table 1 there is a significant excess of DR3/4 heterozygotes. The same type of observation has also been made for CD. From this we conclude that there must be some kind of interaction between the two disease susceptibility genes. One way how this interaction could function is the formation of hybrid molecules (1). Since the DR chains are not polymorphic, the interest focusses on DQ where both α and β-chains are polymorphic so that trans-hybrid molecules can form (2). Such hybrid molecules may be less functional than molecules formed by chains coded for in cis position and they might therefore convey disease susceptibility.

V. HAPLOTYPE SHARING IN MULTIPLE CASE FAMILIES

Investigations in multiple case families, particularly in affected sib pairs have shown, that also for pairs who are not DR3/4 heterozygotes, there is a significant excess of sibpairs sharing both HLA haplotypes. Similar findings were made also in Coeliac Disease and Juvenile Chronic Arthritis. These data indicate, that - at least in a great majority of cases - both HLA-haplotypes contribute to disease susceptibility.

VI. PROTECTIVE ALLELES

It has been observed that the antigen DR2 is very infrequent in patients with Juvenile Diabetes and therefore the speculation has been raised that DR2 could have a "protective" effect. Obviously a simply decreased frequency for DR2 (as compared to controls) is not enough to postulate a protective effect because in a patient-population where alleles like DR3 and DR4 are so strongly increased in frequency all other alleles must be decreased. If one assumes that all non DR3/4 alleles are neutral with respect to disease susceptibility, one would expect that the decrease should be distributed evenly among these alleles proportional to their relative frequencies in the control population. Thus one can calculate the expected degree of compensatory decrease for each allele and compare these values with the observed frequencies. If one performs such analyses with a number of sufficiently large samples of patients with Juvenile Diabetes one finds that indeed the frequency for DR2 is in almost all studies significantly lower than one would expect on the basis of the above described compensatory decrease (Table 2). Nearly the same is true for DR5. Therefore it is justified to ascribe a "protective" effect to the presence of DR2 and of DR5. On the other hand one can also observe that DR1 occurs in a frequency which is much <u>higher</u> than one would expect. Thus DR1 may be termed "permissive" with respect to disease suscep-

tibility to Juvenile Diabetes and in general it can be stated that in Juvenile Diabetes the majority of the known DR alleles (DR1, DR2, DR3, DR4, DR5) are not neutral with respect to disease susceptibility for Juvenile Diabetes.
If one performs such calculations in large bodies of data assembled for other disease studies, one can find that the presence of DR2 is not only protective in Juvenile Diabetes but also in Coeliac Disease as well as in Rheumatoid Arthritis and in Juvenile Chronic Arthritis, while on the other hand DR2 is strongly associated with the susceptibility for Multiple Sclerosis and even more so with the susceptibility for Narcolepsy. As one can see in Table 3, in the different diseases there are various patterns of associated D-alleles, permissive D-alleles and protective D-alleles.
We are quite intrigued by the observation of Sasazuki et al (3), who found an immune suppression gene within the HLA-DQ-region on the DR2 haplotype. We have earlier observed an association of low leucocyte responsiveness to inducers of interferon with HLA-DR2 (4). It appears appealing to us, that this as well as Sasazuki's findings and the attribution of protective effects against disease susceptibility by the DR2 haplotypes might be an expression of the same biological phenomenon.

REFERENCES

1. S. Scholz and E. Albert (1983): Immunological Rev. Vol. 70: 77-88.
2. D.J. Charron, V. Lotteau and P. Turmel (1984), Histocomp. Testing 1984:539-544.
3. T. Sasazuki, S. Matsushita, K. Hirayama, A. Kimura, I. Kikuchi, Y. Nishimura, K. Tsukamoto, M. Yasunami, M. Muto, T. Sone and T. Hirose (1987): New Approach to Genetic Diseases (T. Sasazuki Ed.) Academic Press, Tokyo. 89-97.
4. J. Abb, H. Zander, H. Abb, E. Albert and F. Deinhardt (1983): Immunology 49:239.
5. A. Cambon-de Mouzon, E. Ohayon, G. Hauptmann, A. Sevin, M. Abbal, E. Sommer, H. Vergnes and J. Ducos (1982): Tissue Antigens 19:366-379.
6. Mouzon-Cambon de A., Gilbert J.C., Darnaud J., Abbal M., Segonds A., Sevin A., Denard Y. and Ohayon E. (1981): Ref.Fr.Endocrinol.Clin.Nutr.Metabol. 22:25-33.
7. Brautbar Ch., Karp M., Amar A., Cohen I., Cohen O., Sharon R., Topper E., Levene C., Cohen T. and Albert E.D. (1981): Human Immunol. 3:1:1-13.
8. R. McKenna, E.D. Albert and C.F. McCarthy (1984): Irish Journ.Med.Science 153:7:238-241.

9. Amoroso (1983) personal communication.
10. Bertrams J. and M.P. Baur (1984): Histocomp.Test. 1984, 348-358.
11. Rotter J.I., Anderson C.E., Rubin R., Congleton J.E., Terasaki P.I and Rimoin D. (1983 a): Diabetes 32: 169-174.
12. P. Stastny (1980): Histocomp.Test. 1980:681-686
13. B.M. Ansell and E.D. Albert (1984): Histocomp.Test. 1984: 368-375.
14. Dworschak M. (1982): Doctoral thesis, Univ. Munich.
15. Scholz S. (1985): unpublished own data.

TABLE I: The DR3/DR4 Heterozygote Effect in Various Studies

	DR3/4	DR3/3	DR4/4
French Basques (N = 51) Cambon et al 1982	16	18	1
Toulouse (N = 129) Cambon et al 1981	14	30	4
Israel (N = 92) Brautbar et al 1982	36	5	18
Ireland (N = 49) McKenna et al 1982	22	5	4
Torino (N = 72) Amoroso et al 1983	13	12	6
Munich (N = 90) Scholz et al 1983	22	4	11
9th Workshop (N = 167) Bertrams et al 1984	56	12	13
Combined Data (N = 772) Rotter et al 1983	254	54	50
Total (N = 1422)	433	140	107

TABLE II: Involvement of HLA-DR Alleles in Juvenile Diabetes in Various Populations

	DR1	DR2	DR5	DRw6	DR7	DRw9
Workshop 84	++	(-)	-	/	(+)	/
Workshop 80	/	--	--	NT	-	/
Basque	(+)	(-)	(-)	/	/	/
Toulouse	++	--	--	/	+	/
Spain	(+)	/	(-)	/	/	/
Italy	NT	(-)	-	NT	NT	NT
Munich	(+)	-	/	/	(+)	/
Austria	+	/	(-)	(+)	(+)	/
Israel	/	/	(-)	/	+	/
Ireland	(-)	-	/	(	(+)	/
Japan 80	(+)	--	/	NT	+	+
Orientals 84	(+)	--	(-)	(-)	/	+
Chinese	NT	(-)	NT	NT	NT	+
Overall Trend	+	-	-	/	(+)	+ Ori

* primary association
\+ "permissive"
\- protective
(-) protective, but not significant
/ no deviation
NT not tested

TABLE III: Pattern of Susceptibility and Protection Associated with HLA-DR.

	DR1	DR2	DR3	DR4	DR5	DRw6	DR7	DRw8
Juvenile Diabetes Combined Data N = 1422 Data of Table 2	+	-	*	*	-	/	(+)	/
Rheumatoid Arthr. Workshop 1980 N = 329 Stastny et al 80	+	-	/	*	-	-	-	/
Juv. Chron. Arthr. Workshop 1984 N = 143 Ansell et al 84	(+)	-	+	-	*	+	-	*
Coeliac Disease N = 120 Dworschak et al 82	/	-	*	/	+	(+)	*	NT
Multiple Sclerosis N = 181 Scholz et al 1985	/	*	+	/	(+)	(-)	/	(*)

* primary association
\+ "permissive"
\- protective
(-) protective, but not significant
/ no deviation
NT not tested

HLA-LINKED IMMUNE SUPPRESSION GENE MAPS WITHIN THE HLA-DQ SUBREGION

Takehiko Sasazuki[1], Sho Matsushita, Kenji Hirayama, Akinori Kimura, Ikuo Kikuchi, Yasuharu Nishimura, Kikuo Tsukamoto, Michio Yasunami, Masahiko Muto, Toshio Sone, and Tadaaki Hirose.

Department of Genetics
Medical Institute of Bioregulation
Kyushu University
Fukuoka, JAPAN

I. INTRODUCTION

In foregoing studies, it became clear that the HLA-linked low or nonresponsiveness to streptococcal cell wall (SCW) antigen is controlled by the antigen-specific suppressor T cells, as a single dominant trait(1). We therefore designated the gene as the immune suppression gene for SCW antigen (Is-SCW). Although we attempted to map Is-SCW, analyzing 23 Japanese families, a precise mapping was not feasible because there was no crossover family between HLA genes and Is-SCW. The studies using schistosomal japonica (Sj) and SCW antigens elucidated that the two haplotypes (HLA-DR2-Dw2-DQw1 and HLA-DR2-Dw12-DQw1) are clearly distinguishable in their association with immune responsiveness to Sj and SCW antigens(2). Namely, the HLA-Dw2 haplotype is associated with low responsiveness to SCW whereas the HLA-Dw12 is a high responder haplotype. On the contrary, Dw2 is a high responder haplotype to Sj whereas Dw12 is associated with a low response to Sj. Based on these

[1]Supported by Grants-in Aid 59870019 and 60480117 from the Ministry of Education Culture and Science, Japan.

observations, we compared the two haplotypes at the protein and DNA levels and found that HLA-DQ molecules count significantly in their difference. We also tested the effect of anti HLA-DQ monoclonal antibodies on low responsiveness to SCW or Sj antigen.

II. EVIDENCE FOR HLA-LINKED IMMUNE SUPPRESSION GENES

A. Immune Response to Streptococcal Cell Wall Antigen

To test the immune response to SCW antigen, peripheral blood lymphocytes (PBL, $1x10^5$) from healthy individuals were cultured with 0.5 µg/ml of SCW for 7 days. Distribution of the immune responsiveness to SCW antigen expressed by ln cpm using 113 unrelated healthy adults showed a clear bimodality with a cut-off point of 9.6(3). One hundred and ten members of 23 families were designated as either low or high responders to SCW antigen, by using the same cut-off point, and the low responsiveness proved to be a dominant trait, by the maximum likelihood method (3). The close linkage between HLA and immune responsiveness to SCW antigen was revealed (lod score = 4.311 at recombination fraction = 0.00) by sequential linkage test (3). Proliferative T cell response to SCW antigen required monocytes and the responding T cells belonged to CD 8^-4^+ T cells. T cells from high responders showed marked proliferative responses to SCW antigen, in the presence of allogeneic monocytes from low responders who shared at least one HLA-DR antigen with the T-cell donors. T cells from low responders, on the other hand, never showed an immune response even in the presence of HLA-DR shared high responder monocytes (1). These observations indicate that the HLA-linked low responsiveness to SCW antigen is determined by T cells, not by monocytes. The CD 8^+4^- T cells from low responders completely suppressed the T cell response of high responders. Furthermore, low responder PBL showed a significant immune response to SCW antigen, if $CD8^+4^-$ T cells were depleted, indicating that the low responsiveness to SCW antigen was controlled by antigen-specific suppressor T cells and that the HLA-linked gene controlled the low responsiveness through $CD8^+4^-$ suppressor T cells. We therefore designated this HLA-linked gene controlling the low responsiveness to SCW antigen as HLA-linked immune suppression gene (Is-gene) (1,4). Population analysis revealed that the frequency of this HLA-linked gene was 0.235 ± 0.048 and that this gene was in linkage disequilibrium with HLA-DQw1 (d=0.0223, t=2.2), but not with any alleles in DP or class I loci (5).

B. Immune Response to Schistosomal Japonicum Antigen

We observed that the individuals infected with Sj could be classed into high and low responders at the ratio 1:4.7, using the T cell response to the Sj antigen in vitro. The low responders to the Sj antigen had a strong association with HLA-Bw52-DR2-Dw12-DQw1 haplotype. The $CD8^{+}4^{-}$ suppressor T cells existed in low responders and completely suppressed the immune response of PBL from the high responders to the Sj antigen. This suppression was antigen-specific, and PBL from low responders did respond to the Sj antigen after eliminating the suppressor T cells, thereby suggesting the existence of an HLA-linked Is-gene for Sj antigen (6). Ninety-four percent of the patients with post-schistosomal liver cirrhosis were high responders. All these observations indicate that the low responsiveness to the schistosomal antigen controlled by the HLA-linked Is-gene via the $CD8^{+}4^{-}$ suppressor T cells prevents development of post-schistosomal liver cirrhosis. The high responsiveness to the schistosomal antigen due to the lack of the HLA-linked Is-gene in the homozygous state thus predisposes the individual to post-schistosomal liver cirrhosis.

C. Immune Response to Cryptomeria Pollen Antigen

We carried out a genetic analysis of cryptomeria pollinosis (CP), the most common type I allergy caused by cryptomeria pollen in Japan. Studies on 165 multiplex families revealed that the resistence to CP is an HLA-linked single dominant trait (3). The resistance to CP was in negative linkage disequilibrium with HLA-DQw3 ($p < 0.03$).

Table I. HLA-linked immune suppression genes

Antigen	Response measured by	Linkage with	Linkage disequilibrium	Response is
SCW	T cell proliferation	HLA	DR2, DR5, DQw1	Recessive
CP	IgE response	HLA	DQw3(negative)	Recessive
Sj	T cell proliferation	HLA	Dw12	(Recessive)

Moreover, the healthy individuals had $CD8^{+}4^{-}$ T cells which suppress the autologous response of B + Monocytes + $CD8^{-}4^{+}$ T cells, in an antigen specific manner. Thus we designated this HLA-linked gene for Is-CPAg(3,7). The HLA-linked Is-genes we have clarified are summarized in Table I.

III. DIFFERENCE BETWEEN HLA-Dw2 AND HLA-Dw12 HAPLOTYPE AT THE PROTEIN LEVEL

Although we attempted to map HLA-linked Is-genes by finding a crossing-over between immune suppression and HLA antigens, none was detected. We then tried to pinpoint the gene, comparing HLA-DR2-Dw2-DQw1 and HLA-DR2-Dw12-DQw1 which show different responses to the two antigens. The linkage disequilibrium with the immune responsiveness to SCW antigen and Sj antigen is summarized in Table II.

Table II. The immune responses of Dw2 and Dw12 haplotypes.

Haplotypes	Immune response to	
	SCW	Sj
DR2-Dw2-DQw1	Low	High
DR2-Dw12-DQw1	High	Low

B lymphoblastoid cell lines from HLA-Dw2 and Dw12 homozygotes were radiolabelled with L-(^{35}S) methionine and the cell lysate was immunoprecipitated using monoclonal antibodies to HLA-DR, DQ and DP molecules, then analyzed on two dimensional gel electrophoresis (8). As shown in Figure 1, α chain and one of two β chains (β^1) precipitated with anti DR monoclonal antibody were identical between Dw2 and Dw12, while the other DR β chain was not. Both DQ α^1 and DQ β^1 chains precipitated with anti DQ monoclonal antibody were different between Dw2 and Dw12. Anti DP monoclonal antibody precipitated DP β^1 which is probably identical between the two haplotypes and possibly DP α^1 chain which was not distinguishable from the DQ α^1 chain. Taken together, Dw2 and Dw12 haplotypes are identical in their DR α, DR β^1 and DP β^1 chains, and are nonidentical in their DR β^2, DQ α^1 and DQ β^1, and possibly DP α^1

chains. All these data indicate that the difference in immune responsiveness between Dw2 and Dw12 haplotypes to SCW antigen and Sj antigen could be derived from either DR($\alpha\beta^2$), DQ($\alpha^1\beta^1$), or DP($\alpha^1\beta^1$) molecules.

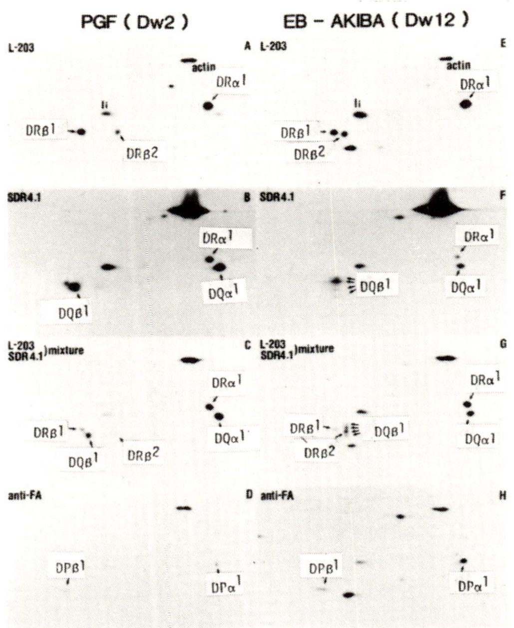

Figure 1. 2D gel analysis of the immunorecipitates from Dw2 or Dw12 homozygous B lymphoblastoid cell line (PGF and EB-Akiba, respectively) with anti DR framework monoclonal antibody L-203 (used in A,C,E and G), anti DQw1 monoclonal antibody SDR4.1 (in B, C, F and G), or with anti FA monoclonal antibody (in D and H).

IV. EFFECT OF ANTI HLA MONOCLONAL ANTIBODIES ON THE LOW RESPONSIVENESS

A. Low Responsiveness to SCW Antigen

We found that anti HLA-DR monoclonal antibodies blocked

the immune response to SCW antigen, Sj antigen and CPAg(9). The interaction between T cells and monocytes to respond to these antigens is restricted by the HLA-DR antigen, indicating that the HLA-DR antigens act as products of immune response genes to respond to these antigens. Based on the observation that the monoclonal antibody to a product of the immune response gene blocks the response, it is conceivable that a monoclonal antibody to a product of immune suppression gene blocks the immune suppression. When PBL from low responders to SCW antigen were cultured with SCW antigen in the presence of various monoclonal antibodies to HLA, the anti HLA-DQ monoclonal antibody restored the immune response to SCW antigen, whereas monoclonal antibodies to DP, class I or control ascites did not affect the low response as shown in Figure 2.

The PBL from 12 low responders and 8 high responders were tested using the same procedure and the "restoration index" was calculated. As shown in Table III, the index in the HLA-DQw1 (+) low responders was significantly high compared with that of DQw1(-) low responders or high responders.

Of three molecules which differ between Dw2 haplotype and Dw12 haplotype ($DR\alpha\beta^2$, $DQ\alpha^1\beta^1$ and $DP\alpha^1\beta^1$), the $DR\alpha\beta2$ molecule seems to act as a product of the immune response gene, because it acted as a restriction molecule in cooperation between the SCW antigen-specific T cell line and monocytes to respond to SCW antigen (8). Anti DP monoclonal antibody did not affect the responsiveness of low responders. Taken collectively, the HLA-DQ molecule is most likely to be a product of Is-SCW.

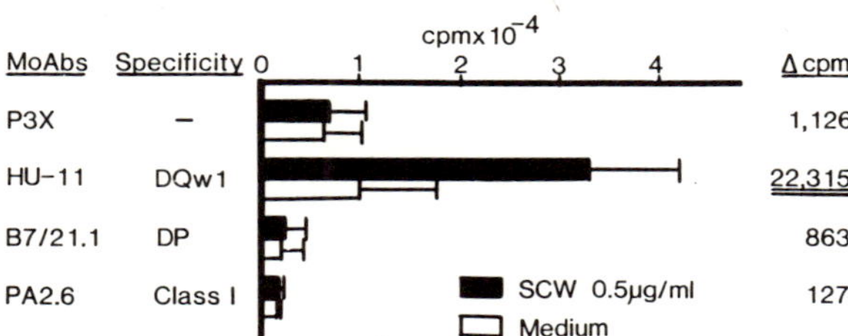

Figure 2. Restoration of the proliferative response to SCW. PBL from a low responder with HLA-A2,-, B27, 35, Cw1, 3.1, DR1,2, DQw1 were cultured in the presence of 0.5μg/ml of SCW antigen for 7 days. Ascites form of monoclonal antibody or control ascites were added to the culture at the final dilution of 1/1000.

Table III. The effect of anti HLA-DQw1 monoclonal antibody on the immune response to SCW antigen.

PBL donor	DQw1	Number of panels tested	Restoration index* SCW 0.5μg/ml	SCW 5.0μg/ml
Low responder	+	9	2.40±1.97	3.44±2.83
	–	3	0.86±0.61	1.40±0.59
High responder	+	6	0.65±0.50	0.72±0.21
	–	3	0.54±0.09	0.60±0.43

PBL from low or high responders to SCW antigen were cultured, with or without anti HLA-DQw1 monoclonal antibody, in the presence of SCW antigen (0.5μg/ml or 5.0μg/ml).
* Restoration index = cpm with MAb/ cpm without MAb

Table IV. The effect of anti HLA monoclonal antibodies on the immune response to Sj antigen.

PBL donor	MAb	Specificity	Immune response to Sj antigen	SCW antigen
SH	Exp 1			
			cpm	cpm
Nonresponder	(–)		3438	55199
to Sj antigen	HU-11	DQw1 +	14596	40466
High responder	SDR4.1	DQw1	23336	89368
to SCW antigen	HU-4	DR framework	962	16933
DR2, w13				
Dw12, w19	Exp 2			
DQw1, –				
	(–)		1301	17143
	HU-11	DQw1 +	8739	23084
	SDR4.1	DQw1	5331	6612
	HU-4	DR framework	1468	478
	HU-18	DQw3	23	26618

PBL from donor SH who showed a high response to SCW antigen and a low response to Sj antigen were cultured with SCW antigen or Sj antigen, in the presence of monoclonal antibodies at a final concentration of 1:100.

B. Low Responsiveness to Sj Antigen

PBL from low responders to Sj antigen were also tested in terms of the restoration of the specific immune response by anti HLA-DQ monoclonal antibody. As shown in Table IV, using PBL from donor SH, who showed a low response to Sj antigen and high response to SCW antigen, the restoration of the response to Sj antigen was observed when anti HLA-DQw1 monoclonal antibody was added to the culture. On the contrary, the immune response to SCW antigen was not affected.

V. DIFFERENCE BETWEEN HLA-Dw2 AND HLA-Dw12 HAPLOTYPE AT THE DNA LEVEL

In order to investigate the molecular difference of DQw1 genes between Dw2 and Dw12 responsible for the difference in immune responsiveness to SCW antigen and Sj antigen between these two haplotypes, we isolated the DQw1β cDNA from a consanguineous HLA-Dw12 homozygous cell line (EB-AKIBA) using three synthetic oligonucleotide probes, the sequence of which was deduced from the known amino acid sequence of the DQw1 chain of Dw2 haplotype (10). The whole base sequence of the DQw1β cDNA from Dw12 was determined and translated into amino acid sequence. A comparative study of the amino acid sequence of DQw1β1 chain of Dw12 with that of Dw2 haplotype revealed that there was a 11 amino acid substitution within the β1 & β2 domains. Out of 11 amino acid substitution, 10 were located within the β1 domain, especially around the 70th residue from the N-terminal, thereby constructing a hypervariable region (data not shown). This may well explain the difference in immune responsiveness between Dw2 and Dw12 haplotypes, namely, the function of the Is-gene.

VI. DISCUSSION

We clearly demonstrated that the HLA-linked immune suppression gene (Is-gene) for SCW antigen is mapped within the HLA-DQ subregion through 1) the comparison of Dw2 and Dw12 haplotypes both in protein and DNA level and 2) the restoration of immune response by anti HLA-DQ monoclonal antibody.

The HLA-linked Is-genes seem to play important roles in the pathogenesis of certain multifactorial disorders. In case of schistosomal infection, a high response to Sj antigen, which can be attributed to Is(-) homozygosity, leads to post-schistosomal liver cirrhosis. In cryptomeria pollinosis, Is-CPAg controls the resistance to cryptomeria pollinosis. Whether the HLA-linked Is-SCW has any significant role in the pathogenesis of streptococcus-related diseases can be investigated from a similar standpoint.

ACKNOWLEDGMENTS

We are grateful to Dr. M. Aizawa of Hokkaido University, Sapporo, Japan, Dr. F. Bach of University Minnesota, Minneapolis, USA, Dr. P. Parham of Stanford University, USA, Dr. P. Wernet, Medical University Clinic, Tübingen, Federal Republic of Germany, Dr. L.A. Lampson, University of Pennsylvania, USA, and Dr. W.F. Bodmer, Imperial Cancer Research Fund Laboratories, England, for providing monoclonal antibodies. We also thank Dr. H. Festenstein, London Hospital Medical College, England, for providing B lymphoblastoid cell line, PGF.

REFERENCES

1. Nishimura, Y and Sasazuki, T. Nature 302:67 (1983).
2. Ohta, N., Nishimura, Y., Iuchi, M. and Sasazuki, T. Clin. exp. Immunol. 49:493 (1982).
3. Sasazuki, T., Nishimura, Y., Muto, M. and Ohta, N. Immunol Rev. 70:51 (1983).
4. Sasazuki, T., Kaneoka, H., Nishimura, Y., Kaneoka, R., Hayama, M. and Ohkuni, H. J. Exp. Med. 152:297s (1980).
5. Nishimura, Y., Sasazuki, T., Amamiya K. and Hirosawa, K. in Recent Advances in Streptococci and Streptococcal Diseases. Y. Kimura, S. Kotami and Y. Shiokawa ed. Reedbooks, Berks. p266. 1985.
6. Ohta, N., Minai M. and Sasazuki, T. J. Immunol.131:2524 (1983).
7. Matsushita, S., Muto, M., Suemura, M, Saito, Y. and Sasazuki, T. J. Immunol. 138:109 (1987).
8 Sone, T., Tsukamoto, K., Hirayama, K., Nishimura, Y., Takenouchi, T., Aizawa, M. and Sasazuki, T. J. Immunol. 135:1288 (1985).
9. Sasazuki, T., Nishimura, Y., Kikuchi, I., Hirayama, K., Tsukamoto, K., Yasunami, M., Matsushita, S., Muto, M., Sone T. and Hirose, T. in Regulation of Immune Genes. 1986. M.Feldman and A. McMichael ed. Humana press, New Jersey. p197.
10.H. Goetz et al. Hoppe-Seyler's Z. Physiol. Chem. 364:749 (1983).

FOREIGN GENE EXPRESSION IN TRANSGENIC MOUSE AND ITS USE FOR THE ANALYSIS OF GENETIC DISORDERS

Ken-ichi Yamamura

Institute for Medical Genetics
Kumamoto University Medical School
Kumamoto, Japan

I. INTRODUCTION

Genetic variants whether naturally occurring or experimentally induced are valuable for determining the physiological role of gene products. These variants are also useful as animal models. However, it is very difficult to identify and characterize the gene that is responsible for abnormal phenotypes. One way to overcome such a problem is to introduce cloned genes into fertilized mouse eggs (1). With this approach, introduced genes may be integrated into mouse chromosomes at an early stage of development, and embryos then develop into adult mice. Using this approach model animals may be produced. In case of genetic disorders which show dominant inheritance, animal models may be made by simply introducing cloned mutant genes into mouse embryos. However, in the case of recessive disorders we still do not have an established technique. There are three possible approaches to make an animal model of recessive disorders. First is to obtain insertion mutation (2). And once such a mutation occurs, the mutated gene can be cloned easily by using the injected gene as a probe. But the integration site

1. Supported by a grant and grant-in-aid for special project research from the Ministry of Education, Science, and Culture of Japan.

is at random and so this approach is of limited use. The second approach is to introduce anti-sense RNA genes. This approach is clearly effective in inhibiting the gene expression in the cell culture system (3). This may be due to the strong positive selection under which more than one hundred times anti-sense RNA can be produced. However this approach does not seem to work in transgenic mice because no positive selection can be done. As an alternative to this approach, we are investigating whether antibodies can neutralize the gene product. So we first analyzed the control of the expression of immunoglobulin gene, whether the introduced immunoglobulin gene is expressed to form complete immunoglobulin molecules, and whether there is any harmful effect within the cell.

A. Requirement of the Enhancer Sequence in the Tissue-specific Expression of the Immunoglobulin Gene

Gene transfer experiments, in which cloned rearranged wild-type or in vitro mutated Ig genes have been introduced into terminally differentiated lymphoid and nonlymphoid cells, have demonstrated that the enhancer sequence located in the intron between the J segment and C region controls the tissue-specific expression of Ig genes (4-6). In addition, Grosshedl and Baltimore (7) recently demonstrated two new regulatory regions that provide transcriptional tissue specificity.

We analyzed the role of the enhancer sequence for the expression of the Ig gene in transgenic mice. We have produced transgenic mice by microinjecting either humanγ 1 Ig gene with (HIG1) (Fig. 1a) or without (VCEγ1) (Fig. 1b) intact enhancer sequence.

HIG1 gene was isolated from human plasma cell leukemia cell line, ARH-77. This gene was shown to contain all the sequences necessary for the expression in the B cell line (8).

We have produced transgenic mice by microinjecting HIG1 gene (9). Tissue specificity of human γ 1 gene expression was examined by fluorescence staining of histological sections using anti-human IgG specific antibody. Fluorescence-positive cells were found only in the spleen of transgenic mice. We then analyzed whether human γ 1 genes were expressed in B or T lymphocytes. Total RNA were isolated from spleen before and after treatment with either

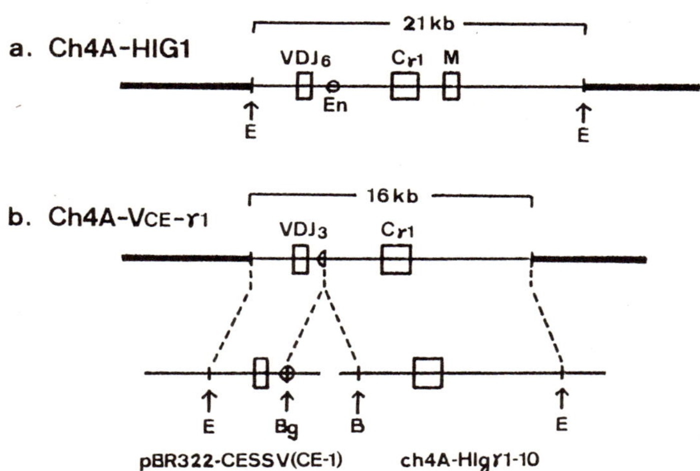

FIGURE 1. Structure of the human γ 1 immunoglobulin gene. V, variable region; D, diversity region; J, joining region; C, constant region; M, membrane exon; E, EcoRI; Bg, BglII; B, BamHI.

lipopolysaccharide (LPS) or concanavalin-A (Con-A) for 3 days and were analyzed by Northern blotting. The level of transcription was increased only when the spleen cells were treated with LPS suggesting that the human γ 1 gene was expressed in B lymphocytes but not in T lymphocytes.

This was also confirmed at the protein level. Spleen cells were stained with anti-human IgG specific antibody before and after treatment with either LPS or Con-A. Human γ chain production was induced by LPS but not by Con-A.

VCE-γ1 gene was constructed by ligating an EcoRI-BglII fragment of pBR322-CESSV (CE-1) containing rearranged VDJ region with a BamHI-EcoRI fragment of Ch4A-Hig 1-10 containing 1 constant region. Human enhancer sequence is known to be located about 1kb down stream from the last JH segment and the Bgl II site is located on this region. Thus by treating CESSV (CE-1) with Bgl II part of the enhancer sequence can be deleted. Five transgenic mice were obtained by injecting VCE-γ1 into fertilized mouse eggs (11). Spleen cells were analyzed for the production of human chain before and after treatment with LPS. But no production of human γ chain was observed in all these transgenic mice.

From these data, the enhancer sequence may be necessary for efficient expression of human γ 1 immunoglobulin gene in transgenic mice.

B. Formation and Excretion of the Complete IgG Molecule

In order to analyze whether or not the human γ chain is associated with endogenous mouse light chains, the molecular weight of bands which react with anti-human IgG specific antibody was examined by Western blotting analysis. As the molecular weight of 1 bands was about 140,000, the human γ chain appeared to be coupled with the mouse light chain to form a complete IgG molecule.

Human γ chain concentration in culture supernatant was assayed by enzyme-linked immunoassay (ELISA). Human γ chain was detected after spleen cells were stimulated with LPS suggesting that human γ chain is secreted into the cell culture medium.

C. Effect of human γ chain production on mouse endogenous Ig gene expression

The production of mouse endogenous Ig in transgenic mouse was examined by immuno-fluorescence staining of spleen cells before and after LPS stimulation. The production of mouse endogenous Ig appeared to be the same as in normal mice. This was confirmed by double-staining experiments. No B lymphocytes, which produce only human γ chain were found, although B lymphocytes that produce only mouse endogenous Ig chain or produce both mouse endogenous Ig chain and human γ chain were found. Thus these results suggest that the expression of human γ chain did not prevent rearrangement or expression of mouse endogenous Ig heavy chain genes.

Ritchie et al. (12) showed that expression of the introduced kappa gene and association of its light chain products with a heavy chain prevents rearrangement of the endogenous kappa genes. Conversely, in those hybridomas in which the foreign kappa gene was inactive (which was a surprisingly large fraction) or unassociated with a functional heavy chain, the endogenous kappa genes did rearrange to produce a functional immunoglobulin. Rusconi and Kohler (13), who introduced both μ and κ genes into mice, observed partial suppression of endogenous immunoglobulin gene rearrangement. Weaver et al. also (14) found that in cells from transgenic mice which express the μ heavy chain gene, a substantial proportion of the heavy-chain allele remains in the germ line configuration.

Thus κ and μ chain gene products seem to have an inhibitory effect on endogenous immunoglobulin gene expression. In our case, the level of human γ chain expression may be too low to inhibit the rearrangement of endogenous mouse immunoglobulin genes.

ACKNOWLEDGMENTS

This work was done in collaboration with A. Kudo, K. Araki, T. Watanabe, H. Kikutani, N. Takahashi, T. Taga, S. Akira, K. Kawai, K. Fukuchi, Y. Kumahara, T. Honjo, and T. Kishimoto.

REFERENCES

1. Gordon, J.W., Scangos, G.A., Plotkin, D.J., Barbosa, J.A. and Ruddle, F.H. (1980). Proc. Natl. Acad. Sci. USA.77:7380.
2. Jaenisch, R., Breindl, M., Harbers, K., Jahner, D. and Lohher, J. (1985). Cold Spring Harbor Symp. Quant. Biol. 50:439.
3. Kim, S.K. and Wold, B.J. (1985). Cell 42:129.
4. Gillies, S.D., Morrison, L.L., Oi, V.T. and Tonegawa, S. (1983). Cell 33:717.
5. Banerfi, J., Olson, L. and Schaffner, W. (1983). Cell 33:729.
6. Queen, C. and Baltimore, D. (1983). Cell 33:741.
7. Grosschedl, R., Weaver, D., Baltimore, D. and Costantini, F. (1984). Cell 38:647.
8. Kudo, A., Ishihara, T., Nishimura, Y. and Watanabe, T. (1985). Gene 33:181.
9. Yamamura, K., Kudao, A., Ebihara, T., Kumino, K., Araki K., Kumahara, Y. and Watanage, T. (1986). Proc. Natl. Acad. Sci. USA. 83:2152.
10. Rabbitts, T.H., Foster, A., Baer, R. and Hamlyn, P.H. (1983). Nature 306:806.
11. Yamamura, K., Kikutani, H., Takahashi, N., Taga, T., Akira, S., Kawai K., Fukushi, K., Kumahara, Y., Honjo, T. and Kishimoto, T. (1984). J. Biochem. 96:357.

12. Ritchie, K.A., Brinster, R.L. and Storb, U. (1984). Nature 312:517.
13. Rusconi, S. and Kohler, G. (1985). Nature 314:330.
14. Weaver, D., Costanitini, F., Imanishi-Kari, T. and Baltimore, D. (1985). Cell 42:117.

III. MOLECULAR APPROACH

MUSCULAR DYSTROPHY AND CYSTIC FIBROSIS

N.J. LENCH, G. BELL, M. FARRALL, P.J. SCAMBLER, B.J. WAINWRIGHT, E. WATSON and R. WILLIAMSON

St. Mary's Hospital Medical School, University of London, London W2 1PG, England.

For diseases where the biochemical defect is known, such as phenylketonuria or the haemoglobinopathies, it is possible to understand the molecular pathology by cloning the mutated gene. This approach is not possible for either Duchenne muscular dystrophy (DMD) or cystic fibrosis (CF), since the biochemical lesion is unknown - as is the case for approximately 80% of inherited diseases caused by a mutation at a single genetic locus. Therefore alternative techniques are required - the study of candidate genes and chromosomal regions, the comparison of cDNA libraries from normal and affected individuals prepared from tissues known to express the defect, and the determination of linkage between random DNA markers and the disease locus itself.

In the case of DMD, localisation of the defect to human chromosome Xp21 was first indicated by the study of dystrophic females with a translocation in this region. Probes from Xp were isolated from an X-specific chromosome library, and there are now over 20 linked probes that can be used for carrier detection and prenatal diagnosis. Localization of the CF locus to 7q22-7q311 was achieved by the help of chromosome-specific saturating linkage mapping with regionally assigned gene probes, following the confirmation of linkage to the unassigned enzyme marker paraoxanase. There are now at least six linked markers, two of which can be used for carrier detection and prenatal diagnosis in informative families.

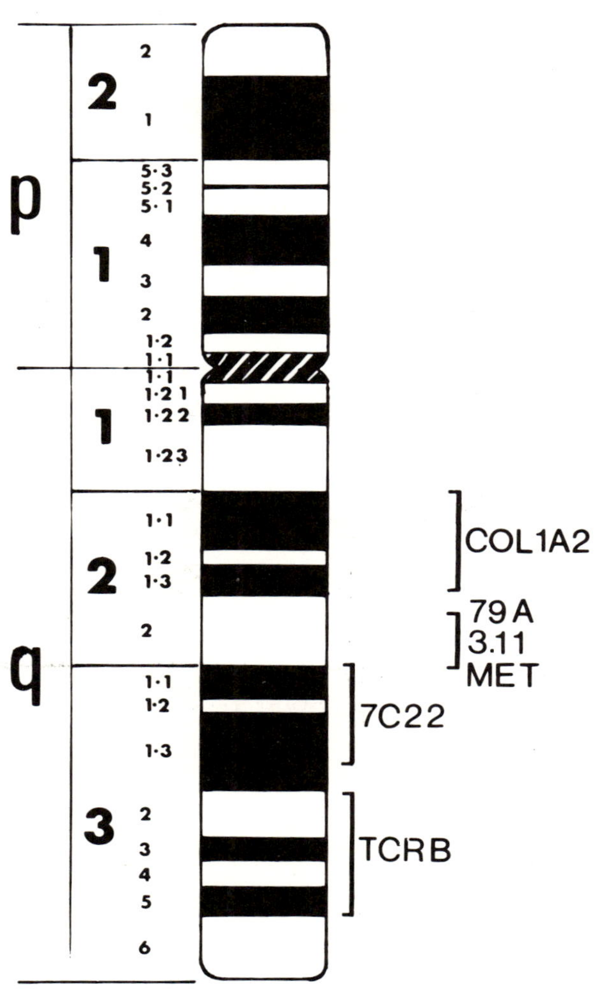

MAPPING OF CHROMOSOME 7

CLINICAL FEATURES - CYSTIC FIBROSIS

Cystic fibrosis is an autosomal recessive disease which has an incidence of 1/1600 in the British population, and a carrier frequency of approximately 1/20. It is the most common severe single-gene inherited disease in Northern Europe. Most cases present as a "failure to thrive" at ages between six months and five years, although some affected infants are detected at birth due to compacted meconium ileus. CF is confirmed by the diagnostic sign that sweat sodium and chloride concentrations are two to three times greater than the normal range (Goodchild and Dodge, 1984).

The major clinical symptoms are due to insufficient pancreatic secretion (which has to be corrected by supplementing the diet with pancreatic enzyme extracts) and viscous mucus which leads to lung infection. Regular and vigorous physiotherapy reduces the likelihood of lung infection, but ultimately antibiotic resistant bacteria colonize the lungs, often the species *Pseudomonas aeruginosa*. Few CF patients survive beyond the age of thirty years even in countries where clinical care is excellent, such as Australia. Carriers of the CF gene show no known symptoms or heterozygote advantage and at present there is no method available for heterozygote detection in the general population.

CLINICAL FEATURES - MUSCULAR DYSTROPHY

Duchenne muscular dystrophy is the most common X-linked lethal disease in man, affecting approximately one boy in 2000. The patients are wheelchair bound by the age 8-10 and generally do not live past the age of 25 years. No known biochemical defect has been identified although numerous theories have been put forward. The hypothesis that there is a structural and/or functional defect of the muscular plasma membrane is the most attractive. This would explain both the leakage of muscle enzymes, such as creatine kinase, into the serum of patients and carriers, as well as the excess calcium uptake by dystrophic muscle fibres. Some female DMD carriers may be identified by elevated serum creatine kinase levels, although this is only diagnostic in approximately 70% of cases (Moser, 1984).

HUMAN MOLECULAR GENETICS

There are three techniques presently being used to study the basic genetic defect causing diseases such as CF or DMD. Although they will be described separately, their usefulness is obviously enhanced when considered as complementing one another.

Candidate genes: a candidate gene is defined as one in which there is a biochemical, genetic or epidemiological reason to believe that a mutation of the DNA sequence might cause or contribute to the genetic component or clinical phenotype of the disease. Sometimes a candidate gene is obvious - it was surmised that a mutation in the gene for phenylalanine hydroxylase is the cause of phenylketonuria long before it was proven by cloning the gene (Woo et al., 1983). Sometimes the argument for candidature is less direct, as in the case of the insulin receptor in juvenile onset diabetes.

Over the years there have been a large number of suggestions as to which genes might be the basic defect causing CF. Among these are superoxide dismutase, calmodulin, albumin (Alon and Riordan, 1984) and complement 3. (Davies et al., 1983c). If a gene probe is available for any candidate gene, then it is relatively simple to test this hypothesis if the gene reveals a restriction fragment length polymorphism (RFLP) and informative families with multiple affected sibs are available.

In the case of Complement 3, for instance, the gene has been cloned and polymorphisms identified (Whitehead et al., 1982); for any family with multiple affected sibs each sib must inherit the same allele from a heterozygote parent since recombination is extremely unlikely within a coding gene (less than 0.001% for each meiosis). One such family that was studied had three affected sibs, each of whom had a different genotype. Therefore, assuming CF to be a single locus disease, this single family excludes C3 as the mutated gene. This provides a powerful tool in excluding genes because the result is unequivocal. Several other candidate genes have also been excluded in this way.

Candidate Regions of Chromosomes: Chromosomal regions can also be treated in the same way as candidate genes. This was most effectively demonstrated in the isolation of markers shown to be linked to DMD. Twelve females were known to show a Duchenne-like muscular dystrophy phenotype associated with a translocation of the X chromosome with a breakpoint at or near Xp21, (Elejalde and Elejalde, 1983), from which it was surmised that this region is implicated in the pathogenesis of DMD. Probes from this region were then isolated from an X-specific chromosome library and shown to be linked to the DMD locus. (Davies et al., 1981; Murray et al., 1982). Once a mutation for a disease has been localized, that region becomes a new candidate region for any other genetic diseases that are related variants, as was the case with Becker muscular dystrophy (Kingston et al., 1983).

HUMAN GENE MAPPING FOR LINKAGE STUDIES

DNA sequence variants arise from single base changes, insertions or deletions. This can often alter the number and distribution of restriction sites in the sequence of DNA in question and they are thus said to display restriction fragment length polymorphisms (RFLPs). The first demonstration of the usefulness of a random RFLP (that is, one which is not related functionally to the gene under study) in inheritance studies was by Kan and Dozy (1978), who followed a point mutation which eliminates a HpaI site 5 kilobases to the 5'-side of the β-globin gene in many genomes of West Africans, in linkage disequilibrium with the gene for sickle β-globin. Shortly after, we made a similar observation for β-thalassaemia (Little et al., 1980). After Kan and Dozy's paper appeared Solomon and Bodmer (1979) were quick to point out that it was theoretically possible to construct a complete linkage map of the human genome using a small number of selectively neutral informative DNA markers exhibiting an RFLP. This concept, later extended by Botstein (1980), has turned out to be accurate.

The applicability of RFLPs in genetic mapping depends on how common they are in the human genome. Jeffreys (1979), on his studies of the β-globin gene locus, has demonstrated that base changes occur on average once in every hundred base pairs. However,polymorphism studies have shown that the distribution of RFLPs throughout the genome is not random. Various calculations have been made on the number of polymorphic DNA markers needed to provide a complete linkage map of the human genome; as few as 200 fully informative DNA probes equally spaced along the human chromosome complement would allow the mapping of any gene to a chromosomal locus using a small number of "affected" families.

This calculation is based upon an ideal situation, and in practice several factors make the real figure for the number of probes required much greater. First, male and female recombination fractions are not equal, female meiosis usually involving a higher degree of crossing over. Second, probes cannot be distributed at equal intervals but will be dispersed randomly along a chromosome. Third, the human genome is divided into chromosomes and recombination is not uniform along their length. Finally, any family will only be fully informative for a proportion of the markers that can be assigned.

At the last Human Gene Mapping Conference (HGM 8, 1985) approximately one thousand markers were reported, the majority of them anonymous probes which are localized chromosomally. Construction of linkage maps of each chromosome is well underway, the study of probe:probe segregation in large pedigrees. The Centre d'Etude Polymorphisme Humain was recently formed in Paris to co-ordinate the construction of a linkage map for the human genome using both DNA and protein markers. Forty large, mainly three generation families are available for the study of probe:probe segregation, and the integration of the results from all the groups participating in the project should greatly increase the speed at which a complete map is obtained.

LINKAGE TO DMD

RC8 was the first probe to be shown to be linked to DMD at a distance of 15 map units (Murray et al., 1982; Davies et al., 1983b), and was localized to Xp21-Xp22.3. A flanking marker L1.28 was subsequently isolated and this was linked at 15-20 map units. If the genetic distance between RC8 and L1.28 is assumed to be 30 map units, then a comparison between chromosome length and genetic length can be made, which predicts that 1 centimorgan is equal to approximately one million base pairs in this region of X chromosome.

These markers, and others which were isolated soon afterwards, are sufficiently close toDMD for recombination to be relatively rare. Therefore, they can be used for exclusion of carrier status in women at risk (Davies et al., 1983a; Harper et al., 1983; Pembrey et al., 1983). However, the individual must be informtive, prefereably for flanking markers, for the test to be useful. It has been estimated that between four and eight two-allele loci of high frequency would be needed on either side of the DMD locus in order to ensure double heterozygosity at the 90% level.

To date there are at least 11 polymorphic loci flanking the DMD locus available for carrier detection. These have also been used for antenatal diagnosis where the carrier mother is heterozygous for two markers proximal and one distal to the DMD locus. Analysis of the segregation pattern could exclude the small possibility of a double recombinant event between the bridging loci. Recombination in families appears to occur at fairly high frequency; this may indicate that the DMD locus is in a "hot spot" for recombination, or could be explained by a high frequency of preexisting rearrangements in this region of the X chromosome.

Twelve female patients have now been characterized that show a balanced X/autosome translocation with a breakpoint at Xp21. Careful cytogenetic examination of the chromosomes suggests that the breakpoint is not in the same place in all patients, indicating that the DMD locus could be highly heterozygous or that there are position effects. Support for the former is given by the fact that the clinically milder Becker dystrophy is localized to the same region of the X chromosome as DMD, and may be allelic (Kingston et al., 1983). In addition DMD shows a very high spontaneous mutation rate compared to other X-chromosome linked diseases such as haemophilia. One translocation breakpoint has now been cloned, which should aid in the elucidation of the structure of the region of the DMD locus, and hopefull lead to the gene itself (Worton et al., 1984).

CYSTIC FIBROSIS

In the case of cystic fibrosis, we attempted to locate the mutation to a single human chromosome by using a similar approach to that which had proven successful for DMD. There were two chromosomal regions that had been proposed in the scientific literature as candidates, chromosome 4 and the tip of the long arm of chromosome 13.

Chromosome 4 had been suggested as the locus of the gene coding for a protein factor causing ciliary dyskinesis. It was thought that the pathophysiology of CF could be explained by abnormal cilial movement in lung epithelial cells, which in turn could be due to a circulating factor in CF serum (Wilson, 1977). The ciliary dyskinesis factor segregated with chromosome 4, amongst other chromosomes, in rodent-human hybrid cell lines. Linkage analysis with several DNA probes located on chromosome 4, however, showed that the CF mutation is not on that chromosome (Scambler et al., 1985).

Chromosome 13qter was a candidate chromosomal region because a family was identified in which there was an unbalanced translocation between chromosomes 6 and 13. The child who is effectively monosomic for the translocated part of the long arm of chromosome 13 was mentally retarded, and also has CF. Since monosomy for the CF locus would give a risk of 1 in 20 of being affected, it was suggested that the CF locus is at chromosome 13q34 (Edwards et al., 1984). The gene for human Factor X is located at 13q34, and we have an RFLP which is revealed by a cloned cDNA coding for this protein. Linkage analysis in eight CF families shows no evidence of co-segregation between CF and the gene for factor X, strongly suggesting that CF is not at 13q34 (Scambler et al., 1986a).

COMPARISON OF GENE LIBRARIES

Most human cell types express approximately 10,000-20,000 different genes as poly[A]+-messenger RNA, and therefore presumably protein. In all autosomal recessive diseases where the biochemical defect is known, a protein is affected, and the gene mutation in turn causes the protein to have an alteration in amino acid sequence, or to be reduced in amount or totally absent. Therefore it is reasonable to assume that in the case of CF the defect will be biochemically apparent as a missing or altered protein in affected tissues, and hence as a messenger RNA which is altered in amount or sequence.

It would appear to be simple to compare mRNA populations to find difference. However, comparison of libraries of expressed genes (cDNA libraries) poses at least four major problems. The first is that many of the sequences that are expressed as mRNA are present in very low copy number, perhaps only one-two copies per cell. Consequently if hybridization techniques are to be used then long periods of time are required for annealing. Second, it is necessary that the cells being compared have an isogenic expression in all cell types other than the diseased cell type. This is unrealistic since individuals differ greatly in (mostly non-pathological) ways, apart from the inherited of a disease such as CF. Third, the techniques are much more suitable when a gene is not expressed due to a deletion, than for a mutation point. We do not know whether CF is due to a deletion or a point mutation, or to either in different individuals. Finally, a tissue that expresses the defect must be available; in CF these tissues are sweat glands, lung epithelium and pancreatic and salivary ducts, none of which are easy to obtain.

Our group spent several years constructing cDNA libraries from lymphocytes from CF and normal patients (Crampton et al., 1980). No differences were detected which were characteristic of CF. On the other hand, comparison of genomic libraries is hindered by the presence of interspersed repetitive elements (Crampton et al., 1981), sheer complexity and the high degree of random differences between normal genomes. Point mutations have been estimated to occur as often as once every 100-200 base pairs. The frequency of small deletions and inversions in the normal genome is unknown. Therefore at this present time the probing of genomic libraries is not a feasible option although new techniques such as hybridization of synthetic oligonucleotide probes to specific coding sequences may alter the situation in the near future (Berent et al., 1985).

LINKAGE TO CF

During the HGM8 Conference several groups working on CF were able to compare and integrate their exclusion data, and at the same time Eiberg's group in Copenhagen reported linkage of the enzyme paraoxanase to CF (Eiberg et al., 1985). Paraoxanase (PON) is found in human serum and is responsible for hydrolysing the insectiside parathion. Its activity can be assayed relatively simply and it is found that persons have either a high or low serum enzyme activity, low activity being transmitted in families as a recessive trait according (more or less) to simple Mendelian rules. Analysis of paraoxanase activity in CF families with multiple affected sibs showed linkage to CF at approximately 15 map units with a lod score of 3.

Unfortunately PON is not at all well characterized; there is no sequence data and was no chromosomal localization at the time linkage was reported. Therefore attention was focussed on those chromosomal regions that had not been excluded either by absence of linkage to CF or to PON. Almost simultaneously three groups isolated probes located on chromosome 7 that were linked to CF. The Toronto group found loose linkage with an anonymous probe, 917 (Tsui et al., 1985); the Salt Lake City group with the oncogene met (White et al., 1985), and our group at St. Mary's with three probes: with COL1A2, TCRB and pJ3.11 (Wainwright et al., 1985; Scambler et al., 1985b).

Of these, met and pJ3.11 are sufficiently tightly linked to permit antenatal diagnosis and carrier detection in informative families. The present map of chromosome 7 with the position of the probes relative to CF is shown in Fig. 1. Other markers have since been isolated. 7C22, an anonymous probe, is linked at approximately four centimorgans (Scambler et al., 1986b) and there are at least three further anonymous probes useful in family studies. No probes show strong allele disequilibrium with CF, although there is evidence for weak disequilibrium with pJ3.11. This suggests that either CF is a very old mutation and has attained equilibrium with neighbouring gene probes or that the mutation has occurred many times during evolution.

It is now possible to use these informative markers in linkage studies in CF families and to determine whether unaffected sibs are carriers or homozygous normals. This is the first time that accurate assessment of carrier status has been possible although the error rate is still approximately 1%. The proportion of carriers to homozygous normals is approximately 2:1, as predicted.

First trimester diagnosis for CF is also possible for informative families using linked DNA probes. We have carried out 15 such tests using fetal DNA prepared from chorionic villi taken transcervically and using MET and pJ3.11 as probes (Farrall et al., 1986). Risk calculations show that the expected false negative and false positive rates are approximately 1% and 2% respectively, for informative nuclear families with one affected sib. Existing probes enable full diagnosis in 90% of couples seeking advice. In half of the remainder diagnosis of one parental mutant chromosome can be made. In rare cases it is even possible to carry out diagnosis in the absence of a living affected sib, if there is an extended family history (Law et al., 1986).

However, there are several reasons why prenatal diagnosis is far from adequate using linkage, even for phase-known families. First, there will be a proportion of families which are uninformative (approximately 5% in this case), or only partially informative. Of those families that are informative a proportion will be unwilling to make use of ante-natal diagnosis. Linkage is of no benefit to patients who already have CF. Finally, there will be little reduction in the total numbers of CF cases since prenatal diagnosis can only be offered to those families that already have at least one affected child, and, in the absence of linkage disequilibrium, linkage does not permit carrier identification. Therefore, it is neccessary to isolate the gene for CF to enable carrier detection in the general population and to identify the basic biochemical defect such that a more rationale treatment can be offered to patients.

REFERENCES

Alan, N., and Riordan, J.R. (1984). In "Cystic Fibrosis: Horizons". (D. Lawson. ed.) p 314 Wiley, Chichester.

Berent, S.L., Mahmoudi, M., Torczynski, R.M., Bragg, P.W., and Bollon, A.P. (1985). Biotechniques. 3:208.

Botstein, D., White, R., Scolnick, M.H., and Davis, R.W. (1980). Am. J. Hum. Genet. 32:314.

Crampton, J.M., Humphries, S.E., Woods, D. and Williamson, R. (1980). Nucl. Acids Res. 9:4895.

Crampton, J.M., Davies, K.E., and Knapp, T.F. (1981). Nucl. Acids Res. 15:3821.

Davies, K.E., Young, B., Elles, R.G., Hill, M.E., and Williamson, R. (1981). Nature. 293:374.

Davies, K.E., Briand, P., Ionasescu, V., Ionasescu, G., Williamson, R., Brown, C., Cavard, C., and Cathelineau, L. (1983). Nucleic Acids Res. 13:155.

Davies, K.E., Pearson, P.L., Harper, P.S., Murray, J.M., O'Brien, T., Sarfarazi, M., and Williamson, R. (1983). Nuc. Acids Res. 11:2303.

Davies, K.E., Gilliam, C., and Williamson, R. (1983). Mol. Biol. Med. 1:185.

Edwards, J.H., Jonasson, J.A. and Blackwell, N.L. (1984). Lancet. i:1020.

Eiberg, H., Mohr, J., Schmiegelow, K., Nielsen, L., and Williamson, R. (1985) Clin. Genet. 28:265.

Elejalde, B.R., and Elejalde, M.M. (1983). In "Cytogenetics of the X chromosome". Second Ed. pp 225-244. Alan R. Liss Inc., New York.

Farrall, M., Rodeck, C.H., Stanier, P., Lissens, W., Watson, E., Lawy, H-Y., Warren, R., Super, M., Scambler, P., Wainwright, B., and Williamson, R. (1986). Lancet. i:1402.

Goodchild, M.C., and Dodge, J.A. (1985). In "Cystic Fibrosis", Second Ed. Bailliere Tindall, Eastbourne.

Harper, P.S., O'Brien, T., Murray, J., Davies, K.E., Pearson, P.L., and Williamson, R. (1983). J. Med. Genet. 20:252.

Jeffreys, A.J. (1979). Cell. 18:1.

Kan, Y.W., and Dozy, A.M. (1978). Proc. Natl. Acad. Sci. USA. 75:5631.

Kingston, H., Harper, P., Pearson, P., Davies, K.E., Williamson, R., and Page, D. (1983). Lancet. 2:1200.

Kingston, H., Thomas, T., Pearson, P., Sarfarazi, M., and Harper., P.S. (1983). J. Med. Genet. 20:255.

Law, H-Y., Stanier, P., Williamson, R., Modell, B., Ward, R.H.T., Petrou, M., Old, J., and Farrall, M. (1986). Submitted Prenatal Diagnosis

Little, P.F.R., Annison, G., Darling, S., Williamson., R., Camba, L., and Modell, B. (1980). Nature, 285:144.

Murray, J.M., Davies, K.E., Harper, P.S., Meredith, L., Mueller, C.R. and Williamson, R. (1982). Nature. 300:69.

Moser, H. (1984). Hum. Genet. 66:17.

Pembrey, M.E., Davies, K.E., Winter, R.M., Elles, R.G., Williamson, R., Fazzoni, T.A., and Walker, C. (1983). Arch. Dis. Child. 59:208.

Scambler, P., Farrall, M., Stanier, P., Bell, G., Ramirez. F., Wainwright., B., Bell, J., Lench, N.J., Kruyer, H., and Williamson, R. (1986). Lancet. ii:1241.

Scambler, P.J., Wainwright, B.J., McGillivray, R., Fung, M., and Williamson, R. (1986). Am. J. Hum. Genet. 4:567.

Scambler, P., Robbins, T., Gilliam, C., Boylston, A., Tippett, P., Williamson, R., and Davies, K.E. (1985). Hum. Genet. 69:250.

Scambler, P.J., Wainwright, B.J., Watson, E., Bates, G., Bell, G., Williamson, R., and Farrall, M. (1986b) Nucl. Acids. Res. 5:1951.

Solomon, E., and Bodmer, W.F. (1979). Lancet. i:293.

Tsui. L-C., Buchwald, M., Barker, D., Braman, J., Knowlton, R., Schumm., J., Eiberg, H., Mohr, J., Kennedy, D., Plasvic, N., Zsiga, M., Markiewicz, D., Akot, G., Brown, V., Helmms, C., Gravius, T., Parker, C., Rediker, K.,and Donis-Keller, H. (1985). Science. 230:1054.

Wainwright, B.J., Scambler, P.J., Schmidtke, J., Watson, E., Law, H-Y., Farrall, M., Cooke, H.J., Eiberg, H., and Williamson, R. (1985). Nature. 318:384.

White, R., Woodward, S., Leppert, M., O'Connell, P., Hoff, M., Herbst, J., Lalouel, J-M., Dean, M., and Vande-Woude, G. (1985). Nature. 318:382.

Whitehead, A.S., Solomon, E., Chambers, S., Bodmer, W.F., Povey, S., and Fey, G. (1982). Proc. Natl. Acad. Sci. USA. 79:5021.

Wilson, G.B., and Fudenberg, H.H. (1977). Nature. 266:463.

Woo, S.L.C., Lidsky, A.S., Guttler, F., Chandra, T., and Robson, K.J.H. (1983). Nature. 306:151.

Worton, R.G., Duff, C., Sylvester, J.E., Schmickel, R.D. and Willard, H.F. (1984). Science. 224:1447.

GENE CONVERSION OF 21-HYDROXYLASE GENE IN CONGENITAL ADRENAL HYPERPLASIA

Fumiki Harada
Akinori Kimura
Takehiko Sasazuki

Department of Genetics
Medical Institute of Bioregulation
Kyushu University
Fukuoka, Japan

I. INTRODUCTION

21-hydroxylase deficiency is a major cause of congenital adrenal hyperplasia (CAH), and one of the most common inborn errors of metabolism. The disease affects about one in 5000 to 15000 births of Caucasian (1), and in 10000 to 20000 births in Japanese (2). These patients can be clinically subdivided into three types of the disease; the salt-wasting form, the simple virilizing form and the nonclassical late-onset form or cryptic variant. The disease is inherited as a monogenic autosomal recessive trait closely linked to HLA (3). There are two 21-hydroxylase genes, 21A and 21B, each located to the 3′ side of one of two genes encoding the fourth component of complement, C4A and C4B, in the class III region of HLA (4,5). The 21B gene is functional, whereas the 21A gene is a pseudogene (6). We performed Southern blot analysis of genomic DNA from Japanese patients with the disease, using the 21-hydroxylase cDNA probe. Some of the affected haplotypes did not have certain restriction fragments corresponding to 21B gene, a result not due to deletion of the 21B gene, but rather to conversion of the 21B gene into the nonfunctional 21A gene. Cloning and sequencing analyses of the 21-hydroxylase genes from a patient was performed to obtain evidence for conversion between the 21-hydroxylase genes.

II. SOUTHERN BLOT ANALYSIS OF 21-HYDROXYLASE GENES IN PATIENTS WITH THE DISEASE

The 21B gene, which includes a 3.7kb Taq I fragment, can be distinguished from the 21A gene that includes a 3.2kb Taq I fragment, on Southern blot analysis using a 21-hydroxylase cDNA probe (pC21/3C) (5). We first performed Southern blot analysis of 12 DNAs from Japanese patients with the disease using Taq I (Fig.1). In DNAs from two patients with salt-wasting CAH (Fig.1-f,h), the 3.7kb fragment was absent and only the 3.2kb fragment was present. In DNAs from two patients with salt-wasting CAH (Fig.1-e,g) and from a patient with simple virilizing CAH (Fig.1-j), the 3.7kb fragment had a reproducibly decreased hybridization intensity compared to the 3.2kb fragment, whereas the two fragment of equal intensity were observed in the other seven patients. These data suggested that six of 21 unrelated affected haplotypes (28.6%) lacked the 3.7kb Taq I fragment. A family study confirmed that the absence of the 3.7kb fragment was segregated in Mendelian fashion (Fig.2).

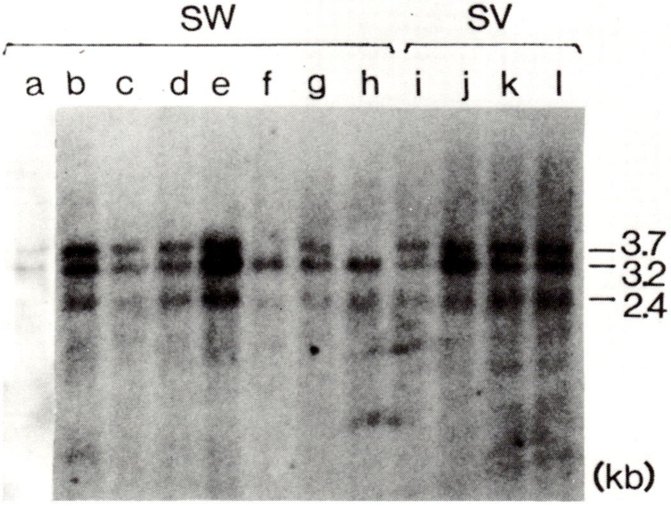

Fig. 1. Southern blot analysis of 21-hydroxylase genes in patients with the disease using Taq I and pC21/3C.
SW : salt-wasting form, SV : simple-virilizing form.

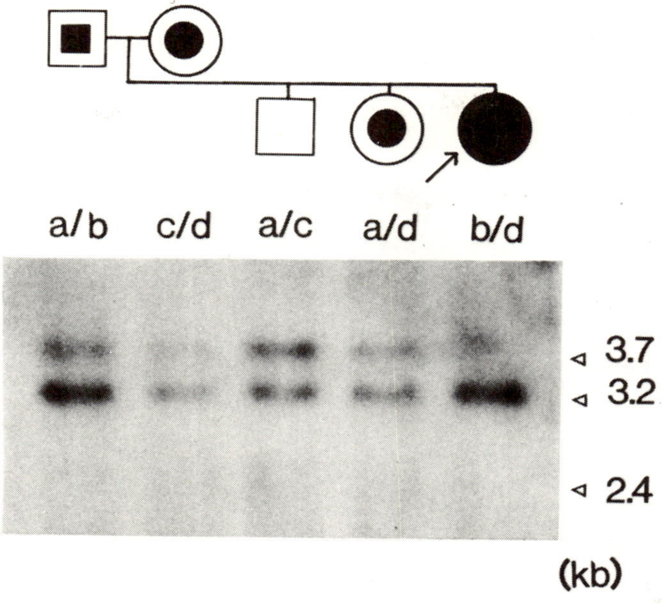

Fig. 2. Segregation of Taq I polymorphism in a family carrying 21-hydroxylase deficiency. The proband is a patient (h) in figure 1.

a : HLA-A31,Bw61,Cw3,DR4
b : HLA-A2, B35, Cw3,DR5
c : HLA-A11, - , - , -
d : HLA-A24,Bw61,- ,DRw9

To understand the nature of the absence of the 3.7kb Taq I fragment, we analyzed the 21-hydroxylase genes of the two patients without the fragment, a normal Japanese individual and hormonally normal Caucasians (AVL,LHM) homozygous for HLA- A1,B8,C4Q*0,C4B1,DR3 haplotype, which has a large deletion that included C4A and 21A genes (7). Southern blot analysis using Pvu II demonstrated that these two patients lacked a 1.7kb Pvu II fragment corresponding to the 21B gene (Fig.2-B). However, Southern blot analysis using Bgl II revealed that these two patients had a 10.5kb Bgl II fragment corresponding to the 21B gene (Fig.2-C). These Southern blot data are summerized in figure 4. All these data suggest that the 21B gene was not deleted, but rather converted into the 21A gene in these two patients.

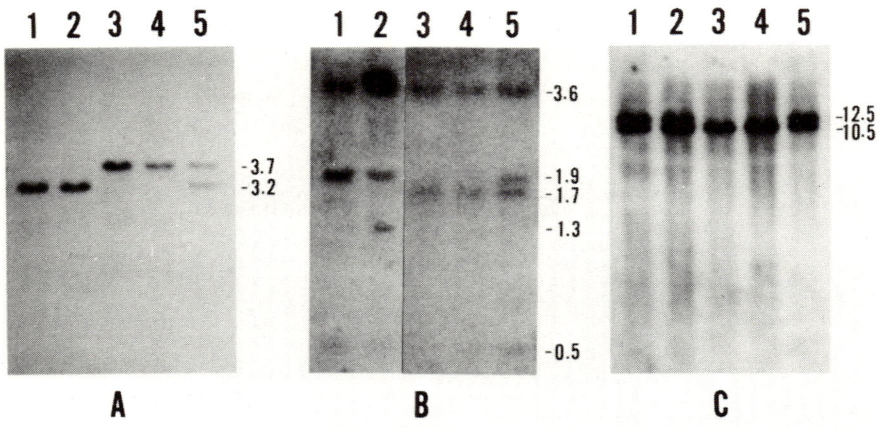

Fig.3. Southern blot analysis of 21-hydroxylase gene from these two patients with salt-wasting CAH and hormonally normal individuals, using Taq I (A), Pvu II (B) and Bgl II (C). The order of genomic DNA samples is : lane 1 (patient (f) homozygous by descent for HLA- A26,B39,C4A3,C4B1,DR4 , lane 2 (patient (h) with HLA-A2,B35,DR5/A2,Bw61,DRw9), lane 3,4 (Caucasian individuals homozygous for HLA-A1,B8,DR3, AVL and LHM, respectively) and lane5 (normal Japanese individual).

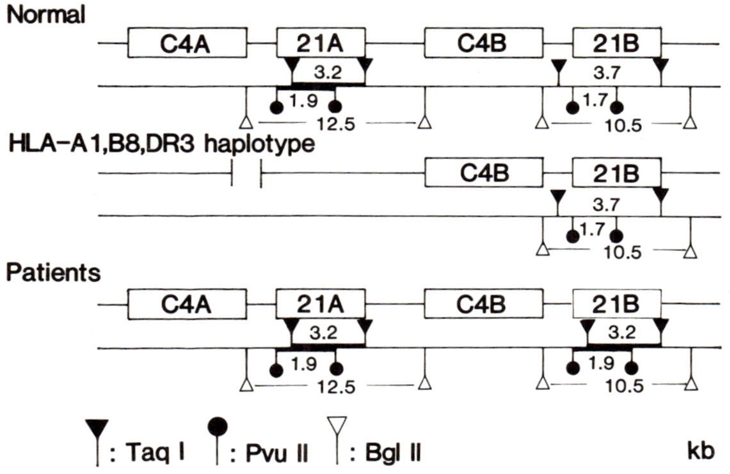

Fig. 4. Schema of Southern blot analysis in figure 3.

II. CLONING AND SEQUENCING ANALYSIS OF 21-HYDROXYLASE GENES IN A PATIENT

In order to obtain direct evidence for the gene conversion, we constructed a human genomic library from one of the two patients homozygous by descent for HLA haplotype and isolated four different clones from 10^6 recombinant phage clones using 21-hydroxylase cDNA probe and C4 cDNA probe (pAT-F)(8). A molecular map of these clones is shown in figure 5.

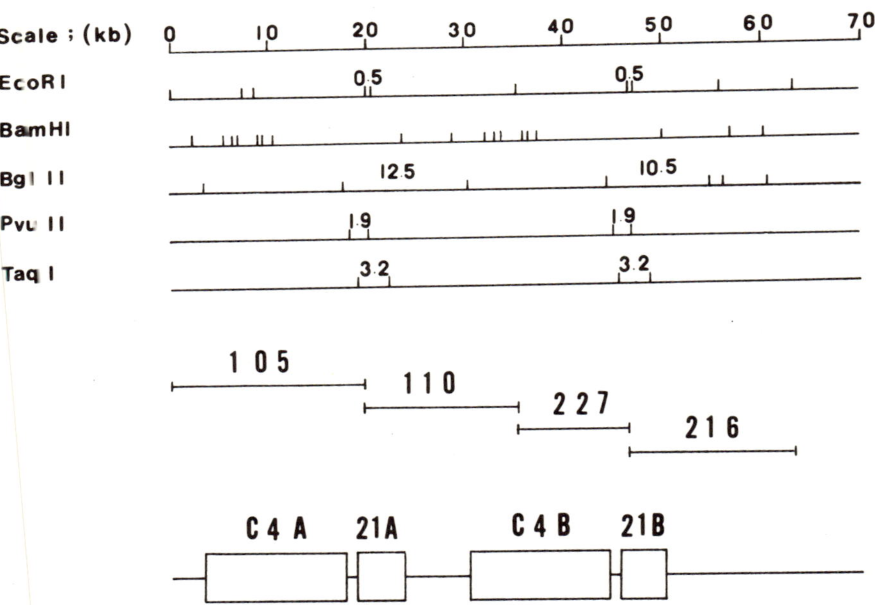

Fig. 5. Molecular map of HLA region containing genes for C4 and 21-hydroxylase in a patient with salt-wasting CAH. The patient is homozygous by descent for HLA -A26,B39,C4A3, C4B1,DR4 and has neither the 3.7kb Taq I nor the 1.7kb Pvu II fragment as demonstrated in Fig. 3- lanel. The 0.5kb EcoRI fragments of clone 110 and that of clone 216 were hybridized to pC21/3C and the nucleotide sequence were determined.

Clone 105 and 227 contained a 2.0 kb EcoRI-Bgl II fragment which hybridized to the C4 probe and weakly to the 21-hydroxylase probe. Clone 110 contained a 0.5kb EcoRI fragment and a 10.0kb EcoRI-Bgl II fragment, both of which hybridized to the 21-hydroxylase probe. Clone 216 contained a 0.5kb EcoRI fragment and a 8.0kb EcoRI-Bgl II fragment, both of which hybridized to the 21-hydroxylase probe. These restriction mapping and hybridization analyses of these clones revealed that the affected haplotype carried two 21-hydroxylase genes, one was a 21A gene in the 12.5 kb Bgl II fragment and another was a 21B gene in the 10.5kb Bgl II fragment. Recently, Higashi et al reported that two 21-hydroxylase genes were highly homologous, but several mutations render 21A gene nonfunctional (6). We subcloned the 0.5kb EcoRI fragments of these two 21-hydroxylase genes and determined the nucleotide sequences. Several differences including a deletion of 8-bases in the third exon that renders 21A gene nonfunctional existed in the sequences of the 21B gene of this patient, indeed the 21B gene was a pseudogene identical to the 21A gene in the sequences. These observations suggested that the 21B gene was converted into a nonfunctional 21A gene, resulting in the salt-wasting CAH in this patient.

IV. DISCUSSION

Gene conversion has been considered to play an important role in evolution of multigene families and is one of the major mechanisms involved in intraspecies sequence homology (9). It has recently become apparent that gene conversion produced a high diversity of the closely related members of a multigene family, such as major histocompatibility class I gene (10). The 21-hydroxylase gene is a member of the cytochrome P-450 multigene family. DNA sequence data of the genes for the closely related phenobarbital-inducible P-450 isoenzymes suggested that gene conversion represented an important evolutionary mechanism for the generation of isoenzymes (11). Moreover, one 21-hydroxylase gene is closely associated with one C4 gene and these two genes are duplicated as a single set. Both the C4 genes are functional but one of 21-hydroxylase genes is a pseudogene. The presence of duplicated and highly homologous sets of the C4 and 21-hydroxylase genes in humans and mice (12,13), suggests that their DNA sequences have been frequently exchanged by a gene conversion-like mechanism. We found that the 21B gene was converted into nonfunctional 21A gene in some Japanese patients with the disease. Affected HLA haplotypes of these two patients analyzed here were not identical. This observation suggests that the conversion of the 21B gene into the 21A gene had occurred independently in their ancestors and some of other three haplotypes which deleted the 3.7kb Taq I fragment might be also explained by the gene conversion.

It is apparent that one of the duplicated 21-hydroxylase genes became a pseudogene through evolution and this may suggest that the presence of functional 21-hydroxylase genes would be unfavorable for the host because of the relative gene dosage effect. If duplication of C4 gene had selective advantage during the evolution , tightly linked 21-hydroxylase gene was high-jacked and also duplicated as a set with C4 gene, and then would become pseudogene. But the presence of such pseudogene closely linked to functional gene in return caused the disease due to the gene conversion between the two genes. The majority of mutations resulting in 21-hydroxylase deficiency have not been characterized, but one-third are deletions of the 21B gene in Caucasian patients (14). We suggest that gene conversion between 21-hydroxylase genes is a relatively common cause of the disease in Japanese patients and this might also account for in part for a predominant cause of CAH. The gene conversion described above would allow us to have a new insight to mechanism for other human monogenetic disorders, where the related homologous genes reside in tandem array.

ACKOWLEDGMENTS

We thank Dr. K. Shimozawa and Dr. J. Yata (Tokyo Medical and Dental University, Japan) for providing clinical data and samples, Dr. P.C. White and Dr. B. Dupont (Memorial Sloan Kettering Cancer Center, USA) for providing the 21-hydroxylase probe, and Dr. M. C. Carroll and the late Dr. R.R. Porter (Oxford University,UK) for providing the C4 probe.

REFERENCES

1. New, M.I., Levine, L.S., in " Congenital adrenal hyperplasia", Monograph on Endocrinology 26 (F. Gross, M.M. Grumbach, eds.), p.44. Spriger-Verlag, Berlin, 1984
2. Naruse, H., Suzuki, E., Irie, M., Tsuji, A., Takasugi, N., Fukushi, M., Matsuura, N., Shimozawa, K., Annals of the New York Academy of Science 458 (M.I. New, ed), p.103. New York Academy of Science, New York, 1985
3. Dupont, B., Oberfield, S.E., Smithwick, E.M., Lee, T.D., Levine, L.S., Lancet 2:1309 (1977)
4. Carroll, M.C., Cambell, R.D., Porter, R.R., Proc. Natl. Acad. Sci. USA 82:521 (1985)
5. White, P.C., Grossberger, D., Onufer, B.J., Chaplin, D.D., New, M.I., Dupont, B., Strominger, J.L., Proc. Natl. Acad. Sci. USA 82:1089 (1985)
6. Higashi, Y., Yoshioka, H., Yamane, M., Gotoh, O., Fujii-Kuriyama, y., Proc. Natl. Acad. Sci. USA 83:2841 (1986)
7. Carroll, M.C., Palsdottir, A., Belt, K.T., Porter, R.R., EMBO. J. 4:2547 (1985)
8. Belt, K.T., Carroll, M.C., Porter, R.R., Cell 36:907 (1984)
9. Ohta, T., Proc. Natl. Acad. Sci. USA 79:3254 (1982)
10. Hayashida, H., Miyata, T., Proc. Natl. Acad. Sci. USA 80:2671 (1983)
11. Achison, M., Adesnik, M., Proc. Natl. Acad. Sci. USA 83:2300 (1986)
12. Belt, K.T., Yu, C.Y., Carroll, M.C., Porter, R.R., Immunogenetics 21:173 (1985)
13. Parker, K.L., Chaplin, D.D., Wong, M., Seidman, J.G., Smith, J.A., Schimmer, B.P., Proc. Natl. Acad. Sci. USA 82:7860 (1985)
14. Werkmeister, J.M., New, M.I., Dupont, B., White, P.C., Am. J. Hum. Genet. In press, (1986)

GENE STRUCTURE OF HUMAN STEROID 21-HYDROXYLASE

Yujiro Higashi
Hidefumi Yoshioka
Miyuki Yamane
Yoshiaki Fujii-Kuriyama

Department of Biochemistry
Cancer Institute
Tokyo

Ayako Tanae

National Children's Hospital
Tokyo

Introduction

Adrenal steroid 21-hydroxylase (C21-OH) belongs to cytochrome P-450 super-gene family and plays a crucial role in the synthesis of steroid hormones such as cortisol and aldosterone (1). In humans, congenital adrenal hyperplasia is observed in a rather high frequency (approximately 1 in 5000 births) and, thus, is one of the most common inborn errors of metabolism. This disease results from a deficiency in one of the enzymes involved in steroidogenesis. Approximately 95% of the affected cases have been reported to be due to a defect only in C21-OH enzyme with an autosomal recessive trait closely linked to the HLA, major histocompatibility complex (2).

Recent studies using gene cloning technique have shown that there are two C21-OH genes, each located near the 3' end of one of the two C4 genes in a relatively short stretch of human chromosomal DNA (3,4). For the first step to study the molecular basis of this genetic disease, we have cloned the two human C21-OH genes and determined their complete nucleotide sequences (5). In the present paper, we describe that the two genes for C21-OH are highly homologous in nucleotide sequence, but that three critical mutations were found in one of the two genes to render the gene nonfunctional. An implication of these observations in this genetic disorder is discussed.

Cloning of The C21-OH Gene and Its Linkage to The C4 Gene.

A human genomic library (6) was screened with the cloned cDNAs containing the entire coding sequence for bovine C21-OH (7). At least five different clones obtained from 10^6 recombinant phages were finally classified into two independent groups by restriction mapping analysis. The two groups, each represented by the two phage clones, are shown in Fig.1. Inserts of the two recombinant phages, λC21A-1 and λC21A-2, overlapping in part with each other, cover the sequence from 9kb upstream to 12kb downstream of one C21-OH gene. On the other hand, λC21B-1 and λC21B-2 also carried another C21-OH gene accompanied by the 9kb upstream and 4kb downstream flanking sequence.

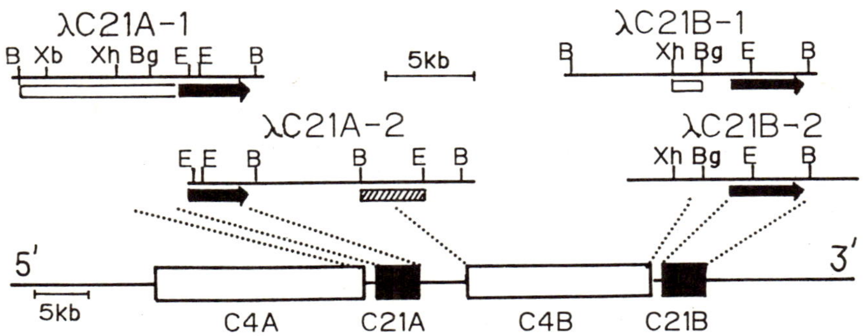

FIGURE 1. Close linkage between C21-OH and C4 genes. Restriction enzyme cleavage maps for λC21A-1 and 2, and λC21B-1 and 2 are shown. Bold arrows indicate the location and direction of the C21-OH genes. Hatched and open boxes show the hybridizing portions with the human C4 5'- and 3'-terminal oligonucleotide probes, respectively. The schematic representation for the tandemly arranged C4 and C21-OH genes was taken from published data (5) and shown at the bottom. Restriction enzyme cleavage sites are indicated as follows, E, EcoRI; B, BamHI; Bg, BglII; Xh, XhoI; Xb, XbaI.

Previous genetic studies of C21-OH deficiency have suggested a linkage of the C21-OH gene to the HLA complex region (2). Recently, close linkage between the human complement C4 and C21-OH genes has been demonstrated by isolating several cosmid clones containing both C4 and C21-OH genes in their single inserts (3,4). Using the C4-specific oligonucleotide probes, we confirmed the linkage of the C21-OH gene to the C4 gene in the insert DNAs of the phage clones we isolated. The results are also shown in Fig.1. Thus, the two C21-OH genes are closely associated with the C4 genes and at least one C21-OH gene, which has been isolated in λC21A-1, seems to be sandwiched between the two C4 genes. From comparison of the restriction maps of our phage clones with those of the cosmid clones reported recently (7), one C21-OH gene isolated in clone λC21A-1 corresponds to the C21-OH "A" gene and the second one in λC21B-1 is equivalent to the C21-OH "B" gene. The characteristic 3.2kb and 3.7kb TaqI fragments (8) are present in clone λC21A-1 and λC21B-1, respectively (Fig.2). The two C21-OH genes alternate with the two C4 genes in a relatively short stretch of chromosomal DNA as shown in Fig.1.

Determination of The Complete Nucleotide Sequences of The Two C21-OH Genes

We determined the complete nucleotide sequences of the C21-OH "A" and "B" genes, which are shown in Fig.2. The sequences for exons and introns were assigned with the aid of homology to the bovine C21-OH cDNA sequence (7) and the GT-AG rule for exon-intron junction. All the exon-intron junctions follow the cannonical GT-AG rule in the C21-OH "A" and "B" genes. The coding nucleotide sequence and the deduced amino acid sequence are approximately 83% and 77% homologous, respectively, between the human gene and the bovine cDNA. A typical TATAA sequence can be seen at the -38 to -34th position and it seems to work as an promoter site from our S1 mapping analysis (Fig.2). At the 486 bp downstream from the termination codon, there exists a typical poly(A) addition signal, AATAAA. By analogy with the bovine C21-OH cDNA sequence, this signal may function as such in the case of humans. On the whole, the human C21-OH gene seems to be approximately 3.4kb long and contains 10 split exonic sequences with very short 9 introns.

From the comparison of the two sequences, we found two important features in the C21-OH gene. One is that the "A" gene is a pseudogene and the other is that the two genes are extremely homologous. We will refer to the two points in the following sections.

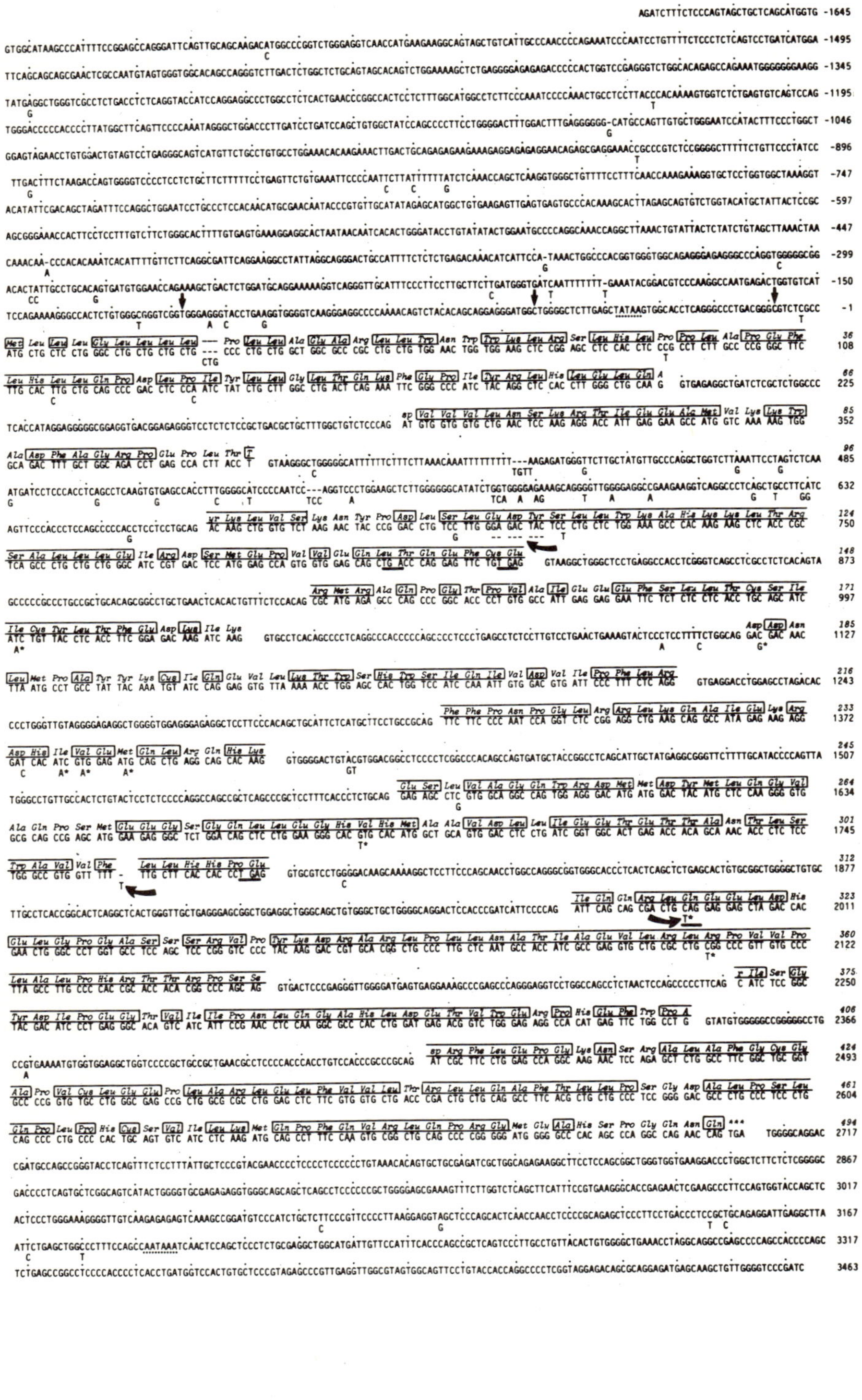

AGATCTTTCTCCCAGTAGCTGCTCAGCATGGTG -1645

GTGGCATAAGCCCATTTTCCGGAGCCAGGGATTCAGTTGCAGCAAGACATGGCCCGGTCTGGGAGGTCAACCATGAAGAAGGCAGTAGCTGTCATTGCCCAACCCCAGAAATCCCAATCCTGTTTTCTCCCTCTCAGTCCTGATCATGGA -1495
C

TTCAGCAGCAGCGAACTCGCCAATGTAGTGGGTGGCACAGCCAGGGTCTTGACTCTGGCTCTGCAGTAGCACAGTCTGGAAAAGCTCTGAGGGGAGAGAGACCCCCACTGGTCCGAGGGTCTGGCACAGAGCCAGAAATGGGGGGGAAGG -1345

TATGAGGCTGGGTCGCCTCTGACCTCTCAGGTACCATCCAGGAGGCCCTGGCCTCTCACTGAACCCGGCCACTCCTCTTTGGCATGGCCTCTTCCCAAATCCCCAAACTGCCTCCTTACCCACAAAAGTGGTCTCTGAGTGTCAGTCCAG -1195
G T

TGGGACCCCCACCCCTTATGGCTTCAGTTCCCCAAATAGGGCTGGACCCTTGATCCTGATCCAGCTGTGGCTATCCAGCCCCTTCCTGGGGACTTTGGACTTTGAGGGGGG-CATGCCAGTTGTGCTGGGAATCCATACTTTCCCTGGCT -1046
G

GGAGTAGAACCTGTGGACTGTAGTCCTGAGGGCAGTCATGTTCTGCCTGTGCCTGGAAACACAAGAAACTTGACTGCAGAGAGAAGAAAGAGGAGAGAGGAACAGAGCGAGGAAACCGCCCGTCTCCGGGGCTTTTTCTGTTCCCTATCC -896
T

TTGACTTTCTAAGACCAGTGGGGTCCCCTCCTCTGCTTCTTTTTCCTGAGTTCTGTGAAATTCCCCAATTCTTATTTTTTATCTCAAACCAGCTCAAGGTGGGCTGTTTTCCTTTCAACCAAAGAAAGGTGCTCCTGGTGGCTAAAGGT -747
G C C G

ACATATTCGACAGCTAGATTTCCAGGCTGGAATCCTGCCCTCCACAACATGCGAACAATACCCGTGTTGCATATAGAGCATGGCTGTGAAGAGTTGAGTGAGTGCCCACAAAGCACTTAGAGCAGTGTCTGGTACATGCTATTACTCCGC -597

AGCGGGAAACCACTTCCTCCTTTGTCTTCTGGGCACTTTTGTGAGTGAAAGGAGGCACTAATAACAATCACACTGGGATACCTGTATATACTGGAATGCCCCAGGCAAACCAGGCTTAAACTGTATTACTCTATCTGTAGCTTAAACTAA -447

CAAACAA-CCCACACAAATCACATTTTGTTCTTCAGGCGATTCAGGAAGGCCTATTAGGCAGGGACTGCCATTTTCTCTCTGAGACAAACATCATTCCA-TAAACTGGCCCACGGTGGGTGGCAGAGGGAGAGGGCCCAGGTGGGGGCGG -299
A G C

ACACTATTGCCTGCACAGTGATGTGGAACCAGAAAGCTGACTCTGGATGCAGGAAAAAGGTCAGGGTTGCATTTCCCTTCCTTGCTTCTTGATGGGTGATCAATTTTTTT-GAAATACGGACGTCCCAAGGCCAATGAGACTGGTGTCAT -150
CC G C T T

TCCAGAAAAGGGCCACTCTGTGGGCGGGTCGGTGGAGGGTACCTGAAGGTGGGGTCAAGGGAGGCCCCAAAACAGTCTACACAGCAGGAGGGATGGCTGGGGCTCTTGAGCTATAAGTGGCACCTCAGGGCCCTGACGGGCGTCTCGCC -1
T A C G T

Met Leu Leu Leu Gly Leu Leu Leu Leu --- Pro Leu Leu Ala Gly Ala Arg Leu Leu Trp Asn Trp Trp Lys Leu Arg Ser Leu His Leu Pro Pro Leu Ala Pro Gly Phe 36
ATG CTG CTC CTG GGC CTG CTG CTG CTG --- CCC CTG CTG GCT GGC GCC CGC CTG CTG TGG AAC TGG TGG AAG CTC CGG AGC CTC CAC CTC CCG CCT CTT GCC CCG GGC TTC 108
CTG T

Leu His Leu Leu Gln Pro Asp Leu Pro Ile Tyr Leu Leu Gly Leu Thr Gln Lys Phe Gly Pro Ile Tyr Arg Leu His Leu Gly Leu Gln A 66
TTG CAC TTG CTG CAG CCC GAC CTC CCA ATC TAT CTG CTT GGC CTG ACT CAG AAA TTC GGG CCC ATC TAC AGG CTC CAC CTT GGG CTG CAA G GTGAGAGGCTGATCTCGCTCTGGCCC 225
C C

sp Val Val Val Leu Asn Ser Lys Arg Thr Ile Glu Glu Ala Met Val Lys Lys Trp 85
TCACCATAGGAGGGGGCGGAGGTGACGGAGAGGGTCCTCTCTCCGCTGACGCTGCTTTGGCTGTCTCCCAG AT GTG GTG GTG CTG AAC TCC AAG AGG ACC ATT GAG GAA GCC ATG GTC AAA AAG TGG 352

Ala Asp Phe Ala Gly Arg Pro Glu Pro Leu Thr T 96
GCA GAC TTT GCT GGC AGA CCT GAG CCA CTT ACC T GTAAGGGCTGGGGGCATTTTTTCTTTCTTAAACAAATTTTTTTTT---AAGAGATGGGTTCTTGCTATGTTGCCCAGGCTGGTCTTAAATTCCTAGTCTCAA 485
C TGTT G G G

ATGATCCTCCCACCTCAGCCTCAAGTGTGAGCCACCTTTGGGGCATCCCCAATCC---AGGTCCCTGGAAGCTCTTGGGGGGCATATCTGGTGGGGAGAAAGCAGGGGTTGGGGAGGCCGAAGAAGGTCAGGCCCTCAGCTGCCTTCATC 632
G G G C T TCC A TCA A AG T A A G T GG

yr Lys Leu Val Ser Lys Asn Tyr Pro Asp Leu Ser Leu Gly Asp Tyr Ser Leu Leu Trp Lys Ala His Lys Lys Leu Thr Arg 124
AGTTCCCACCCTCCAGCCCCACCTCCTCCTGCAG AC AAG CTG GTG TCT AAG AAC TAC CCG GAC CTG TCC TTG GGA GAC TAC TCC CTG CTC TGG AAA GCC CAC AAG AAG CTC ACC CGC 750
G G -- --- --- T

Ser Ala Leu Leu Leu Gly Ile Arg Asp Ser Met Glu Pro Val Val Glu Gln Leu Thr Gln Glu Phe Cys Glu 148
TCA GCC CTG CTG CTG GGC ATC CGT GAC TCC ATG GAG CCA GTG GTG GAG CAG CTG ACC CAG GAG TTC TGT GAG GTAAGGCTGGGCTCCTGAGGCCACCTCGGGTCAGCCTCGCCTCTCACAGTA 873

Arg Met Arg Ala Gln Pro Gly Thr Pro Val Ala Ile Glu Glu Glu Phe Ser Leu Leu Thr Cys Ser Ile 171
GCCCCCGCCCTGCCGCTGCACAGCGGCCTGCTGAACTCACACTGTTTCTCCACAG CGC ATG AGA GCC CAG CCC GGC ACC CCT GTG GCC ATT GAG GAG GAA TTC TCT CTC CTC ACC TGC AGC ATC 997

Ile Cys Tyr Leu Thr Phe Gly Asp Lys Ile Lys Asp Asp Asn 185
ATC TGT TAC CTC ACC TTC GGA GAC AAG ATC AAG GTGCCTCACAGCCCCTCAGGCCCACCCCCAGCCCCTCCCTGAGCCTCTCCTTGTCCTGAACTGAAAGTACTCCCTCCTTTTCTGGCAG GAC GAC AAC 1127
A* A C G*

Leu Met Pro Ala Tyr Tyr Lys Cys Ile Gln Glu Val Leu Lys Thr Trp Ser His Trp Ser Ile Gln Ile Val Asp Val Ile Pro Phe Leu Arg 216
TTA ATG CCT GCC TAT TAC AAA TGT ATC CAG GAG GTG TTA AAA ACC TGG AGC CAC TGG TCC ATC CAA ATT GTG GAC GTG ATT CCC TTT CTC AGG GTGAGGACCTGGAGCCTAGACAC 1243

Phe Phe Pro Asn Pro Gly Leu Arg Arg Leu Lys Gln Ala Ile Glu Lys Arg 233
CCCTGGGTTGTAGGGGAGAGGCTGGGGTGGAGGGAGAGGCTCCTTCCCACAGCTGCATTCTCATGCTTCCTGCCGCAG TTC TTC CCC AAT CCA GGT CTC CGG AGG CTG AAG CAG GCC ATA GAG AAG AGG 1372

Asp His Ile Val Glu Met Gln Leu Arg Gln His Lys 245
GAT CAC ATC GTG GAG ATG CAG CTG AGG CAG CAC AAG GTGGGGACTGTACGTGGACGGCCTCCCCTCGGCCCACAGCCAGTGATGCTACCGGCCTCAGCATTGCTATGAGGCGGGTTCTTTTGCATACCCCAGTTA 1507
C A* A* A* GT

Glu Ser Leu Val Ala Gly Gln Trp Arg Asp Met Met Asp Tyr Met Leu Gln Gly Val 264
TGGGCCTGTTGCCACTCTGTACTCCTCTCCCCAGGCCAGCCGCTCAGCCCGCTCCTTTCACCCTCTGCAG GAG AGC CTC GTG GCA GGC CAG TGG AGG GAC ATG ATG GAC TAC ATG CTC CAA GGG GTG 1634
G

Ala Gln Pro Ser Met Glu Glu Gly Ser Gly Gln Leu Leu Glu Gly His Val His Met Ala Ala Val Asp Leu Leu Ile Gly Gly Thr Glu Thr Thr Ala Asn Thr Leu Ser 301
GCG CAG CCG AGC ATG GAA GAG GGC TCT GGA CAG CTC CTG GAA GGG CAC GTG CAC ATG GCT GCA GTG GAC CTC CTG ATC GGT GGC ACT GAG ACC ACA GCA AAC ACC CTC TCC 1745
T*

Trp Ala Val Val Phe Leu Leu His His Pro Glu 312
TGG GCC GTG GTT TTT - TTG CTT CAC CAC CCT GAG GTGCGTCCTGGGGACAAGCAAAAGGCTCCTTCCCAGCAACCTGGCCAGGGCGGTGGGCACCCTCACTCAGCTCTGAGCACTGTGCGGCTGGGGCTGTGC 1877
T C

Ile Gln Gln Arg Leu Gln Glu Glu Leu Asp His 323
TTGCCTCACCGGCACTCAGGCTCACTGGGTTGCTGAGGGAGCGGCTGGAGGCTGGGCAGCTGTGGGCTGCTGGGGCAGGACTCCACCCGATCATTCCCCAG ATT CAG CAG CGA CTG CAG GAG GAG CTA GAC CAC 2011
T*

Glu Leu Gly Pro Gly Ala Ser Ser Ser Arg Val Pro Tyr Lys Asp Arg Ala Arg Leu Pro Leu Leu Asn Ala Thr Ile Ala Glu Val Leu Arg Leu Arg Pro Val Val Pro 360
GAA CTG GGC CCT GGT GCC TCC AGC TCC CGG GTC CCC TAC AAG GAC CGT GCA CGG CTG CCC TTG CTC AAT GCC ACC ATC GCC GAG GTG CTG CGC CTG CGG CCC GTT GTG CCC 2122
T*

Leu Ala Leu Pro His Arg Thr Thr Arg Pro Ser Se r Ile Ser Gly 375
TTA GCC TTG CCC CAC CGC ACC ACA CGG CCC AGC AG GTGACTCCCGAGGGTTGGGGATGAGTGAGGAAAGCCCGAGCCCAGGGAGGTCCTGGCCAGCCTCTAACTCCAGCCCCCTTCAG C ATC TCC GGC 2250

Tyr Asp Ile Pro Glu Gly Thr Val Ile Ile Pro Asn Leu Gln Gly Ala His Leu Asp Glu Thr Val Trp Glu Arg Pro His Glu Phe Trp Pro A 406
TAC GAC ATC CCT GAG GGC ACA GTC ATC ATT CCG AAC CTC CAA GGC GCC CAC CTG GAT GAG ACG GTC TGG GAG AGG CCA CAT GAG TTC TGG CCT G GTATGTGGGGGCCGGGGGCCTG 2366

sp Arg Phe Leu Glu Pro Gly Lys Asn Ser Arg Ala Leu Ala Phe Gly Cys Gly 424
CCGTGAAAATGTGGTGGAGGCTGGTCCCCGCTGCCGCTGAACGCCTCCCCACCCACCTGTCCACCCGCCCGCAG AT CGC TTC CTG GAG CCA GGC AAG AAC TCC AGA GCT CTG GCC TTC GGC TGC GGT 2493
A

Ala Pro Val Cys Leu Gly Glu Pro Leu Ala Arg Leu Glu Leu Phe Val Val Leu Thr Arg Leu Leu Gln Ala Phe Thr Leu Leu Pro Ser Gly Asp Ala Leu Pro Ser Leu 461
GCC CCG GTG TGC CTG GGC GAG CCG CTG GCG CGC CTG GAG CTC TTC GTG GTG CTG ACC CGA CTG CTG CAG GCC TTC ACG CTG CTG CCC TCC GGG GAC GCC CTG CCC TCC CTG 2604

Gln Pro Leu Pro His Cys Ser Val Ile Leu Lys Met Gln Pro Phe Gln Val Arg Leu Gln Pro Arg Gly Met Gly Ala His Ser Pro Gly Gln Asn Gln *** 494
CAG CCC CTG CCC CAC TGC AGT GTC ATC CTC AAG ATG CAG CCT TTC CAA GTG CGG CTG CAG CCC CGG GGG ATG GGG GCC CAC AGC CCA GGC CAG AAC CAG TGA TGGGGCAGGAC 2717

CGATGCCAGCCGGGTACCTCAGTTTCTCCTTTATTGCTCCCGTACGAACCCCTCCCCTCCCCCCTGTAAACACAGTGCTGCGAGATCGCTGGCAGAGAAGGCTTCCTCCAGCGGCTGGGTGGTGAAGGACCCTGGCTCTTCTCTCGGGGC 2867

GACCCCTCAGTGCTCGGCAGTCATACTGGGGTGCGAGAGAGGTGGGCAGCAGCTCAGCCTCCCCCCGCTGGGGAGCGAAAGTTTCTTGGTCTCAGCTTCATTTCCGTGAAGGGCACCGAGAACTCGAAGCCCTTCCAGTGGTACCAGCTC 3017

ACTCCCTGGGAAAGGGGTTGTCAAGAGAGAGTCAAAGCCGGATGTCCCATCTGCTCTTCCCGTTCCCCTTAAGGAGGTAGCTCCCAGCACTCAACCAACCTCCCCGCAGAGCTCCCTTCCTGACCCTCCGCTGCAGAGGATTGAGGCTTA 3167
C G T C

ATTCTGAGCTGGCCCTTTCCAGCCAATAAATCAACTCCAGCTCCCTCTGCGAGGCTGGCATGATTGTTCCATTTCACCCAGCCGCTCAGTCCCTTGCCTGTTACACTGTGGGGCTGAAACCTAGGCAGGCCGAGCCCCAGCCACCCCAGC 3317
C T

TCTGAGCCGGCCTCCCCACCCCTCACCTGATGGTCCACTGTGCTCCCGTAGAGCCCGTTGAGGTTGGCGTAGTGGCAGTTCCTGTACCACCAGGCCCCTCGGTAGGAGACAGCGCAGGAGATGAGCAAGCTGTTGGGGTCCCGATC 3463

The C21-OH "A" Gene Is a Pseudogene

In the sequence of C21-OH "A" gene (Fig.2), we found several critical nucleotide alterations from the bovine cDNA sequence, which could not be accounted for by the difference in species. These alterations in the "A" gene are (1) an 8 bp deletion in the 3rd exon, (2) a 1 bp insertion in the 7th exon and (3) a transition (C-T) point mutation in the 8th exon (Fig.2, indicated by the arrows). All these mutations presumably render the gene nonfunctional by generating premature terminations. On the other hand, the "B" gene has no such nucleotide alterations. In this gene, the 10 exonic sequences assigned as described in the previous section provide an open reading frame for 494 amino acids, 2 amino acids less than the bovine counterpart. On the basis of these observations, we concluded that the C21-OH "A" gene is a pseudogene, whereas the "B" gene is a genuine gene. This conclusion was indeed substantiated by the RNA blot analysis of an adrenal total RNA preparation using specific probes for the "A" and "B" gene sequences (sequences of the 701st to 720th and 695th to 723rd with a deletion of 8bp and 2 base replacements for the "B" and "A" probe, respectively). Only the probe specific for the "B" gene yielded a hybridization band at about 2.4kb position with the adrenal RNA preparation from a hormonally normal individual (Fig.3).

FIGURE 2. Complete nucleotide sequence and deduced primary structure for the two human C21-OH genes. The complete nucleotide sequence and the deduced primary structure for the C21-OH "B" gene, an intact gene, are shown in a full description. Only nucleotide alterations in the "A" gene, a pseudogene, from the "B" counterpart are indicated under the corresponding nucleotides in the "B" gene. Of these alterations, base substitutions causing putative amino acid changes are marked by asterisks. Gaps represented by bars are introduced to minimize the differences between the two sequences. The three deleterious mutations in the "A" gene are indicated by upward arrows. In-phase termination codons resulting from frame shifts caused by such mutations are underlined. The transcription initiation sites determined by the S1 mapping analysis are indicated by downward arrows. The typical TATAA sequence and the poly(A) addition signal, AATAAA are indicated by dotted lines. Amino acids identical for both human and bovine sequences are also enclosed in boxes.

These results provide the structural basis for the observations reported by White et al (8). They found that deletion of one of the two C21-OH gene was closely associated with the C21-OH deficiency in the patients with HLA-Bw47 haplotype, raising a possibility that the other gene for C21-OH might be nonfunctional.

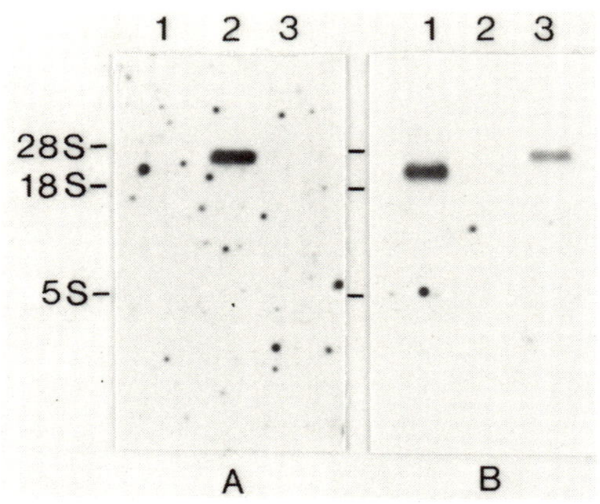

FIGURE 3. Northern blot hybridization analysis of the transcript from the human C21-OH "A" and "B" genes in the adrenal tissue. Two parallel gels were blotted and hybridized with the "A"(A) or "B"(B) gene-specific probe. Lane 1, poly(A) RNA; 2, 3.2kb TaqI fragment of the "A" gene; 3, 3.7kb TaqI fragment of the "B" gene.

The C21-OH "A" and "B" Genes Are Highly Homologous

Except for those critical base alterations, the two C21-OH genes show very extensive sequence homology even in their flanking and intron sequences. In the region of approximately 5.1kb ranging from 1.5kb upstream to 0.6kb downstream of the C21-OH gene, only 88 base alterations were observed between the "A" and "B" genes with an overall sequence homology of 98% (Fig.2).

Recently, White et al. have reported the presence of two C21-OH genes in the mouse genome, each located immediately 3' to the C4 and Slp genes in the H-2 complex region (9) and there also seem to exist the two C21-OH genes in the bovine genome (10). It is reasonable, therefore, to infer that the close association of the C4 and C21-OH genes and their duplication as a unit had occurred before the adaptive radiation of mammals (ca. $7x10^7$ years ago). The extensive homology observed between the human C21-OH "A" and "B" genes could not be expected to have been maintained for such an evolutional time period involved if these genes evolved independently under natural evolutional pressure. Independent evolution of the two intraspecies C21-OH genes would generate approximately 20% sequence divergence between them as observed interspecifically between the human and bovine coding sequences.

Taking such circumstances into considerations, It seems highly probable that sequence divergence between the two human genes have been rectified during this period of time by a mechanism such as concerted evolution involving gene conversion and/or unequal crossing-over, which has been actually observed in class I genes of the MHC complex (11) or in the human fetal γ-globin gene (12), although another possibility cannot be rigorously eliminated, of recent and independent duplication of a set of the C21-OH and C4 genes in the human, murine and probably bovine genomes after the mammalian radiation.

An extensive sequence homology was also reported between the C4 and Slp (C4 equivalent) genes in the mouse genome (13). The presence of the duplicated and highly homologous sets of the C4 and C21-OH genes in a short stretch of chromosomal DNA may allow for frequent exchange of their DNA sequences by gene conversion and/or unequal crossing-over during meiosis.

In these genetic situations, once one of the two C21-OH genes became a pseudogene, the gene conversion or unequal crossing-over might have generated, more often than not, various gene recombination products unfavorable to the host animals. Some of them might have been fixed in the human genome as the affected allelic variant forms of the C21-OH gene. To address this point, we have analyzed the genomic DNAs from the C21-OH deficient patients as well as from the normal individuals by using the Southern blot hybridization analysis.

Southern Blot Hybridization Analysis of The Genomic DNAs from The Patients and The Normal Individuals: An Evidence of Gene Conversion

By the TaqI digestion of the human genomic DNA, the "A" gene, a pseudogene, is included in 3.2kb TaqI fragment and the "B" gene, a functional gene, is in 3.7kb TaqI fragment. The additional 2.6kb band can be observed in Fig.4. However, this is a common flanking TaqI fragment in the two genes which can hybridize with the probe of the genomic DNA fragment we have used.

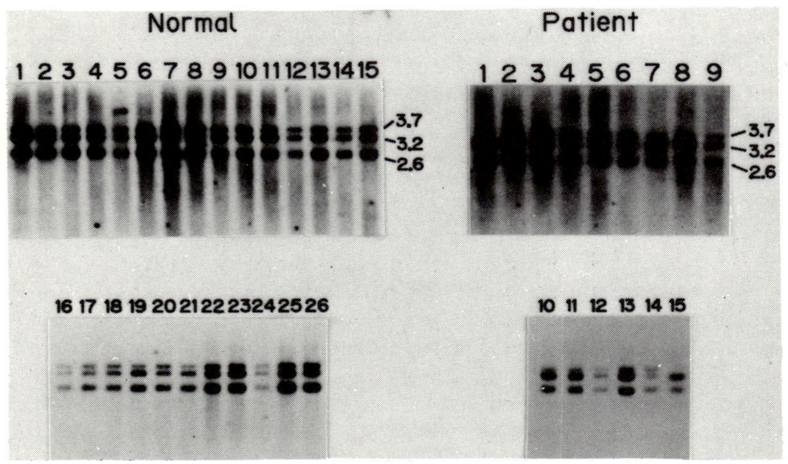

FIGURE 4. Southern blot hybridization analysis of the TaqI-digested genomic DNAs from the patients and the normal individuals. About 2μg of the total DNAs were digested with TaqI and electrophorsed in 0.8% agarose gels. They were blotted and hybridized with ^{32}P-labelled 3.1kb EcoRI-BamHI genomic DNA fragment (Fig.1). Each lane represents an individual genomic DNAs.

In the case of normal individuals, all the genomic DNAs show the 3.7kb TaqI fragment, indicating that all have the functional C21-OH gene. In general, the "A" and "B" genes are estimated to exist in the ratio of 1 to 1 from the intensity of the autoradiogram. There are a few variants such as No.19 and No.25, whose 3.2kb bands are slightly stronger in the intensity of hybridization signal than the 3.7kb band. Interestingly, one individual (No.5) appears to have an extra C21-OH gene, suggesting that the unequal crossing-over event might have occurred in the ancestor of this person.

As for the patient individuals, however, two patients' DNAs (No.1 and No.15) completely lack the 3.7kb sequence, or a functional gene. Similar cases have been reported by White et al. (5). Other patients seem to have the 3.7kb sequence corresponding to the functional gene, although such 3.7kb bands show sometimes less intensities as compared to the normal individuals. This is probably because these patients are often carring the affected haplotype in which the functional C21-OH gene is deleted, or replaced by the pseudogene probably by the mechanisms of unequal crossing-over and/or gene conversion. It has as yet to be clarified whether the 3.7kb sequences in the patients' genomic DNAs are truly functional or not. It is possible that the 3.7kb sequences in the patients' genomic DNAs may have ruinous mutation (point mutation, deletion, insertion and other deleterious mutations), and some of them can be derived from the pseudogene by the mechanism of gene conversion without changing the length of the 3.7kb fragment. Such possibilities will be clarified by the structural analysis of the 3.7kb sequences from the patients.

In at least one patient, however, gene conversion in the 3.7kb sequence appeared to be the case as shown in Fig.5. The "A" gene-specific probe, the same as used in Fig.3, can

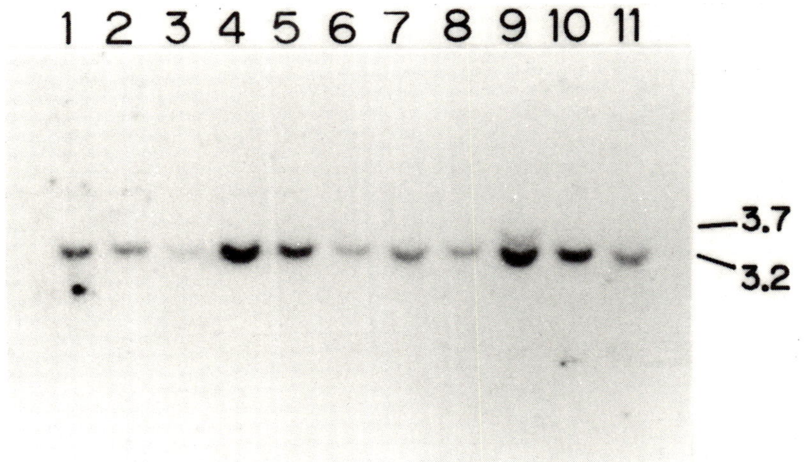

FIGURE 5. Southern blot hybridization analysis of the TaqI-digested genomic DNA from the patients using the "A" gene-specific oliginucleotide probe. Total DNAs were isolated from the EB virus-transformed lymphocyte derived from the peripheral blood lymphocyte of the patients. Hybridization experiments using oligonucleotide probe were carried out as reported previously (14), after DNA was transferred to nitrocellulose filters by blotting. Each lane represents an individual genomic DNA. Patients corresponding to lane 1, 2 and 5 have no 3.7kb "B" gene fragment, but all others have 3.7kb fragment.

hybridize with the 3.7kb TaqI fragment in addition to the usual 3.2kb fragment in the patient No.9. It clearly shows the gene conversion of the 3.7kb sequence to the pseudogene sequence of 3.2kb length. Thus, the 3.7kb sequence has been cnanged to be nonfunctional in this patient and this seems to be the reason for the C21-OH deficiency of this patient.

We are now trying to search other mutations in the 3.7kb sequence using the additional oligonucleotide probes specific for the deleterious mutations found in the pseudogene as shown in Fig.3.

Concluding Comments

Recently, Nathans et al. have reported that the color-blindness, the most frequent genetic disease in human, may result from the unequal recombination or gene conversion between the highly homologous pigment genes (15). The molecular pathogeny of this disease seems very similar to the case of C21-OH deficiency. Since the eukaryotic genome often consists of the multigene family whose constituent genes are tandemly arranged in a short stretch of chromosomal DNA, gene conversion or unequal crossing-over could be one of the potent molecular mechanism to generate the variant forms of haplotype, some of which could be harmful to the host animals.

References

1. Takemori, S. and Kominami, S. (1984) Trends Biol. Sci. 9, 393-396.
2. New, M.I. and Levine, L.S. (1984) in Pediat. adolesc. Endocr., Vol.13, pp.1-46 (eds. New, M.I. and Levine, L.S., Karger, Basel)
3. Carroll, M.C., Campbell., R.D. and Porter, R.R. (1985) Proc. Natl. Acad. Sci. USA 82, 521-525.
4. White, P.C., New, M.I. and Dupont, B. (1984) Proc. Natl. Acad. Sci. USA 81, 7505-7509.
5. Higashi, Y., Yoshioka, H., Yamane, M., Gotoh, O. and Fujii-Kuriyama, Y. (1986) Proc. Natl. Acad. Sci. USA 83, 2841-2845.
6. Lawn, R.M., Fritsch, E.F., Parker, R.C., Blake, G. and Maniatis, T. (1978) Cell 15, 1157-1174.

7. Yoshioka, H., Morohashi, K., Sogawa, K., Yamane, M., Kominami, S., Takemori, S., Okada, Y., Omura, T., and Fujii-Kuriyama, Y. (1986) J. Biol. Chem. 261, 4106-4109.
8. White, P.C., Grossberger, D., Onufer, B.J., Chaplin, D.D., New, M.I., Dupont, B. and Strominger, J.L. (1985) Proc. Natl. Acad. Sci. USA 82, 1089-1093.
9. White, P.C., Chaplin, D.D., Weis, J.H., Dupont, B., New, M.I. and Seidman, J.G. (1984) Nature (London) 312, 465-467.
10. Chung, B.-C., Matteson, K.J. and Miller, W.L. (1985) DNA 4, 211-219.
11. Hood, L., Steinmetz, M. and Malissen, B. (1983) Ann. Rev. Immunol. 1, 529-568.
12. Slightom, J.L., Blechl, A.E. and Smithies, O. (1980) Cell 21, 627-638.
13. Nonaka, M., Takahashi, M., Natsuume-Sakai, S., Nonaka, M., Tanaka, S., Shimizu, A. and Honjo, T. (1984) Proc. Natl. Acad. Sci. USA 81, 6822-6826.
14. Miyada, G.C., Klofelt, C., Reyes, A.A., McLaughlin-Taylor, E. and Wallace, R.B. (1985) Proc. Natl. Acad. Sci. USA 82, 2890-2894.
15. Nathans, J., Piantanida, T.P., Eddy, R.L., Shows, T.B. and Hogness, D.S. (1986) Science 232, 203-210.

AN HLA CLASS I IMMUNE RESPONSE GENE

Andrew McMichael, Frances Gotch, Judy Bastin,
Alain Townsend and Jonathon Rothbard

Nuffield Department of Medicine, University
of Oxford and Imperial Cancer Research Fund
Research Fund Laboratories, London, U.K.

Infectious diseases are not normally considered to be genetic in origin. However the host response to infectious agents undoubtably involves genetic elements and the purpose of this short review is to identify HLA class I molecules as important modifiers of such disease processes. Cell-mediated immune responses play a key role in the host response to infection by viruses and other intracellular parasites(1). While it is clear that antibody has a crucial prophylactic effect and, if specific and neutralising , can offer complete protection, this cannot happen in the first encounter with a virus or where the virus alters its surface glycoproteins. In addition there are instances where the antibody response may facilitate infection and be harmful(2). Circulating cytotoxic T lymphocytes (CTL) have been found in the acute phase of virus infections in humans and in mice(3). Although they may cause disease where the virus infection is not cytopathic(4), in the majority of cases they contribute to recovery by destroying cells that are replicating viruses.

There is extensive evidence in mice that CTL clones transferred to influenza infected recipients will clear the virus whereas antibody does not(3,5). Nude mice cannot be cured of influenza virus infection by antibody but can by transfer of CTL (6); the T cell deficient animals die at a higher frequency and survivors secrete virus indefinitely(7). Cellular immunity in humans may be impaired by malnutrition, drugs and infections or may be congenitally deficient. In these situations, there is an increased susceptibility to virus infection as well as to some fungi and protozoa. In addition, persistent virus infections such those as with Epstein-Barr virus or cytomegalovirus, may be reactivated. The consequences may be fatal.

For any type of infecting virus there are a range of resonses in a previously healthy individual that are partially determined by the virus and partially by the host.

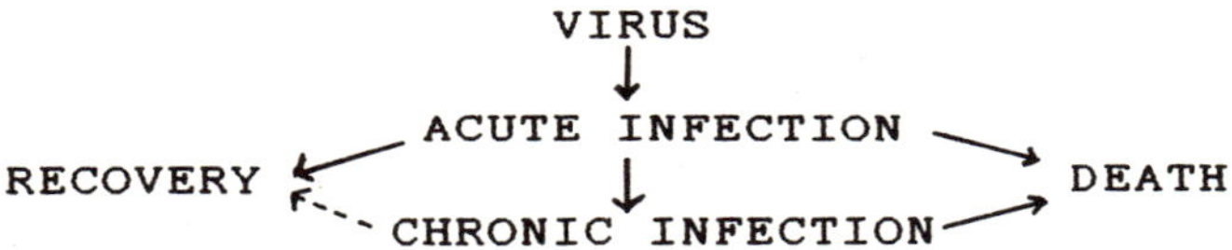

The nature of virulence is poorly understood although a mild virus in one population can appear virulent in another due to poor host defences(8). But even a homogeneous virus strain can be associated with considerable variation in response in a given population. In the influenza pandemic of 1968, for instance, there was a range of responses that varied from asymptomatic infection to death(9).

Given the crucial role of cell-mediated immunity, particularly CTL, in the recovery phase, it is almost certain that the activity of these cells will help to determine the course of an infection. Genetic factors that affect CTL activity must therefore be very important. The existence of immune response genes in the Major Histocompatibility Complex(MHC) that affect T helper cell recognition of foreign proteins on the surface of specialised antigen presenting cells has been known for almost twenty years(10). Evidence for a similar role for Class I MHC antigens has been less forthcoming although these genes are just as polymorphic.

Zinkernagel and Doherty (11) first implicated MHC class I in the murine CTL response to virus infections by showing that CTL recognize virus antigen in association with MHC class I molecules on infected cells.Doherty et al. later showed that H-2b mice only recognized influenza virus in association with Db and not Kb(12); this is now known to be not quite correct because different immunization protocols can induce a weak Kb restricted response(13). Nevertheless this result raised the possibility that immune response gene-like effects existed for class I molecules. Meleif et al(14) showed that, while H-2b mice recognized Sendai virus through Kb and not Db , the Kbm-1 mutant mice could not recognize this virus and were susceptible to fatal infection.

The nature of the involvement of MHC class I antigens in the CTL responses has become much clearer recently from studies on the recognition of influenza virus infected cells . Exactly which virus antigens were seen by CTL was always enigmatic. Polyclonal CTL specific for influenza A virus were shown to cross-react with all A viruses tested but not with influenza B virus(15). This was puzzling, as it was assumed that CTL would recognize the integral membrane glycoproteins of the virus which are abundantly expressed on the surface of infected cells,but which vary considerably between virus strains. When murine CTL were cloned it was found that, although the majority recognised all A viruses, occasional clones recognised virus subtypes.Using viruses with reassorted RNA segments (genes) to infect cells it was possible to map recognition in two cases to the virus genes that code for the internal proteins polymerase PB2 (16) and nucleoprotein (17). These experiments were followed by studies where cDNA copies of mRNA for viral proteins were transfected into L cells and these were then tested as targets with A virus cross-reactive polyclonal CTL (18). It was found that a substantial part of the polyclonal CTL response in mice was indeed directed at the nucleoprotein, an internal virus protein that was not found on the cell surface. Yewdell et al(19) obtained similar results using recombinant vaccinia viruses to insert single influenza gene products into target cells.

We have used the latter method to study the fine specificity of human CTL which are also fully cross-reactive in their recognition of different strains of influenza A virus. The proteins recognised by CTL from the donors studied so far are nucleoprotein(NP), PB2 and matrix-1 (M1)(20). The glycoproteins were not recognised and neither were PB1 PA and NS1 which are known to be recognised by some murine CTL clones.(21). It may be relevant that NP and M-1 are made in relatively large amounts in infected cells and this may bias the response towards these proteins. However matrix-1 protein has not been identified by murine CTL(21).

The CTL response to these proteins was restricted by HLA Class I molecules. When CTL were tested with target cells that shared single HLA A or B products it was found that different virus antigens were associated with different HLA molecules(21). We have repeatedly found that matrix-I is recognised in association with HLA-A2 while NP is poorly recognised with A2 but is seen with other Class I molecules. Thus particular HLA Class I alleles appear to function as immune response genes for single virus proteins.

This association of immune response with HLA types was taken a step further by Townsend, Gotch and Davey(22) who transfected L cells with truncated nucleoprotein cDNA segments. It was found that truncated gene products as short as 130 amino acids (compared to 498 amino acids of the whole protein) could be recognised by influenza specific cytotoxic T lymphocytes from mice. Because the cDNA segments used in these studies were overlapping, it was possible to localize epitopes. C57 (H2b) mice recognised an epitope between amino acids 327 and 386 and, in contrast, CBA CTL (H2k) recognised an epitope between 1 and 130. Although it can not be concluded that this difference is only due to H2 type it is possible that Db selects 327 to 386 and one of the H2k Class I molecules selects 1-130.

The notion that MHC Class I molecules may function as immune response genes was taken a further step forward by the definition of individual epitopes using synthetic peptides. The observations described above had lead to these experiments, in particular the following three points. 1), Internal virus proteins were identified as CTL antigens yet were not identified on the surface of transfected L cells or recombinant vaccinia virus infected cells. 2), In the truncated cDNA experiments it was clear that whole protein was not necessary for recognition, making conformational dependent epitopes unlikely. 3), Because segments 1-130 and 327-386 were both recognised there could be no common leader sequence to take the truncated NP to the cell surface. In addition, monoclonal antibodies to nucleoprotein have repeatedly failed to inhibit CTL recognition of target cells that we know to be NP specific; in contrast anti-H2 or anti-HLA Class I block efficiently.

Synthetic peptides covering the C terminal third of a nucleoprotein were prepared by Bohadur and Rothbard(23). They were added to uninfected histocompatible target cells and tested with CTL preparations. In the first experiments, cloned CTL from C57 mice, specific for an epitope in the 327-386 region (as defined above), lysed target cells treated with the peptide 365-380; ten other peptides were not recognised. Peptide 365-380 was seen in association with H2Db. This region of the NP molecule differs in two amino acids between 1934 and 1968 and both peptides have been synthesised. Polyclonal C57 CTL grown and selected on 1934 virus show specificity for the 1934 peptide and vice versa(19). As it is also possible to obtain C57 CTL that cross-react with both viruses and peptides, the latter must contain at least two epitopes, both of which are restricted through Db.

Following the identification of peptide 365-380 as a recognised peptide by murine CTL, polyclonal human CTL from donor MG, that were known to be NP specific, were tested on the set of peptides. It was found that autologous cells treated with a different peptide,335-349, were lysed(23). The MG CTL recognised this peptide in association with HLA B37.Both the 1934 and 1968 sequence peptides,which differ in a lys-arg at position 348 were recognised. A further twenty NP peptides have been tested on target cells of various HLA type but no other has been recognised by MG CTL,including 365-380.

Influenza A virus specific CTL from fifteen other human CTL donors have been tested for recognition of peptide 335-349. Four others, all of which share HLA B37,recognised this peptide. None of the other eleven identified it and none of these had HLA B37. Two of the five HLA B37 positive individuals were related but the others were not. It seems likely therefore that HLA B37 plays a major role in selecting the NP epitope 335-349 recognised by influenza specific CTL in these individuals. Although we can not completely exclude the possibility that these five individuals uniquely share one or more other polymorphic genes that affect T cell recognition, this seems unlikely.

For the nucleoprotein of influenza virus therefore a number of distinct epitopes have been identified using the synthetic peptide approach. As indicated above there are two in the 365-380 region which associate with H2Db and one between 335-349 which associates with B37. In addition there is an epitope between 1-130, which associates with an H2K Class I antigen. The MHC association is highly specific in each case. The simplest explanation for these results is that peptide binds to the MHC Class I molecule and is recognised by CTL as a complex. Thus each MHC Class I molecule would interact with a series of peptide epitopes, which could of course be derived from many different virus proteins, to stimulate CTL responses.It is possible however that the situation is not so simple and that the MHC Class I molecule plays a role in the selection of the T cell receptor repertoire.

A key question is whether the epitope to which CTL respond makes any difference to the overall

CTL response and hence immunity to the virus. This is clearly true in the extreme example of the failure of H-2Kb ml mice to make a CTL response to Sendai virus. It is possible that if C57 mice were repeatedly challenged, like humans, with a series of related viruses that their NP subtype specificity could give them less cross-reactive protection than in other mouse strains. We have identified humans in whom we are unable to generate an influenza A virus specific CTL response but as we can not challenge them with live virus we do not know whether this non-responsiveness is genetically determined or due simply to low levels of immunity several years after natural infection(24). We would not expect to find many true non-responders in the humans because the population has been exposed to influenza virus for hundreds of years, with many deaths.

The genetic non-responsiveness may be found more easily when a new epidemic virus appears before natural selection occurs. The acquired immune deficiency virus (LAV/HTLVIII/HIV) is a case in point. It would be worth determining whether any HLA antigen confers protection from either acquiring sero-positivity after exposure or developing the lethal immune deficiency syndrome. Such data has not yet been collected.

A final point that is relevant to this symposium, CTL clearly respond to intracellular proteins. Thus attempts at gene therapy may provoke this type of immune response resulting in death of the altered cells. CTL are unlikely to be able to distinguish between virus and non-self human proteins unless immune tolerance has been established for the latter. This is more likely to be a problem when there is no gene product made in the patients. Given the foregoing results HLA type may affect the outcome.

In conclusion, we have found that CTL recognise definable epitopes on virus proteins which are often internal. It is likely that they recognise processed antigen. MHC (HLA) type plays a major role in epitope selection and thus could have an important effect on the quality of an immune response to a virus or other intra-cellular parasites including inserted gene products.

Acknowlegement: This work was supported by the Medical Research Council and the Imperial Cancer Research Fund.

REFERENCES

1. Askonas B.A.,McMichael,A.J.,Webster,R.G. in "Basic and Applied Influenza Research" (A.S.Beare,ed.)p.159.CRC Press Boca Raton 1982
2. Halstead S.B.,Progr.Allergy,31,301,1982.
3. Yap,K.L. Ada,G.L. McKenzie,I.F.C. Nature 273,238,1978
4. Zinkernagel R.M.,Leist,T. Hengartner,H. Althage,A. J.Exp.Med.162,2125,1985
5. Lin,Y.L. Askonas,B.A. J.Exp.Med. 154,225,1981
6. Wells,M.A. Ennis,F.A. Albrecht P J.Immunol 126.1042,1981 .
7. Wells,M.A. Albrecht,P. Ennia,F.A. J.Immunol 126,1036,1981.
8. Barnes R. Papua New Guinea Med.J. 9,127,1966.
9. Foy,H.M. Cooney,M.K.,Allen,I. J.Infect.Dis. 134,362,1976.
10. Benacerraf,B. McDevitt,H.O. Science 175, 273, 1972
11. Doherty,P.C. Blanden,R.V. Zinkernagel,R.M. Transplant.Rev. 29, 89, 1976.
12. Doherty,P.C. Biddison,W.E. Bennink,J.R. Knowles,B.B. J.Exp.Med. 148, 534, 1978
13. Towsend,A.R.M. Taylor,P.M. Mellor,A.L. Askonas,B.A. Immunogen.17, 283, 1983
14. Meleif C.J. in Proceedings of the Sixth IR Gene Workshop,(M.Feldman,ed) in press 1986.
15. Townsend A.R.M. McMichael A.J. Progress in Allergy. 36,10,1985.
16. Bennink J.R.,Yewdell J.W.,Gerhard W. Nature 296,75,1982.
17. Townsend A.R.M. Skehel J.J. J.Exp.Med. 160,552,1984
18. Townsend A.R.M. McMichael A.J. Carter N.P. Huddleston J.A. Brownlee G.G. Cell 39, 13, 1984.
19. Yewdell J.A. Bennink J.R.,Smith G.,Moss B.

Proc.Natl.Acad.Sci.USA.,82.1785,1985.
20. Gotch F.M. McMichael A.J. Smith G.L. Moss B. Submitted 1986.
21. Yewdell J.A. personal communication.
22. Townsend A.R.M. Gotch F.M. Davey J, Cell 42, 457, 1985.
23. Townsend A.R.M. Rothbard J. Gotch F.M. Bahadur G. Wraith D. McMichael A.J. Cell 44, 959, 1986.
24. McMichael,A.J. Gotch,F.M. Dongworth,D.W. Clark,A, Potter,C.W. Lancet 2, 762, 1983.

Maximal Expression of Cloned Human Genes in Cultured Mammalian Cells

Masao Yamada

National Children's Medical Research Center
Tokyo, Japan

I. INTRODUCTION

One approach to the treatment of genetic disease is the replacement of missing, but required, metabolites or gene products. This method has proved successful in the treatment of hemophilia by supplying blood coagulation factors VIII or IX, and in the treatment of congenital deficiencies of hormones, i.e. growth hormone in pituitary dwarfism, thyroid hormone in congenital hypothyroidism, and steroid metabolites in adrenogenital syndromes (Friedmann, 1983). Although it has been possible to obtain small amounts of gene products from autopsy specimens and other natural sources, recent advances in molecular biology have led to the isolation of large amounts of gene products by recombinant DNA techniques and molecular cloning methods. Most works of the overproduction of cloned genes have concentrated on microbial hosts, such as *Escherichia coli* and yeasts, because they can be easily manipulated and cultured in large quantity at low cost (Harris, 1983). However, such host-vector systems are not appropriate for overproducing certain human gene products. The expressed gene products would not be expected to have the posttranscriptional and posttranslational modifications seen in higher eukaryotes. Some genes, especially relatively large genes, are inefficiently expressed in such hosts. Therefore, a system to express cloned human genes in human cells in large amounts is desirable for the broad application of product-replacement to treat genetic diseases.

Many expression systems in cultured mammalian cells have

NEW APPROACH TO
GENETIC DISEASES

been developed (Elder et al., 1981; Rigby, 1982; Gluzman, 1982). They are classified into several categories based on (1) how the cloned genes are introduced into the cultured cells, (2) the vector which assures the replication or integration of the exogenous DNA, and (3) time duration of the expression. Transient expression systems by transfection or microinjection are suitable to examine the activity of cloned genes and their derivatives in order to elucidate the mechanisms of gene regulation. The amount of gene products thus obtained is enough to be analyzed but far less than needed for preparation. Permanent expression systems by integration of the cloned gene into a host chromosome, or by maintenance as an episome, can be used for both analytical and preparative purposes. Viral infection systems in which the cloned gene is introduced into cells after packaging within the virion are the most efficient way to transfer exogenous DNA into a large number of cells at the same time. The more temperate DNA available for transcription, the higher expression is achieved. Thus, the virus infection system is suitable to isolate gene products in large amounts.

II. ADENOVIRUS VECTOR SYSTEM

Adenovirus has several advantages over other virus vector systems for an expression vector. (1) Human adenovirus can infect a variety of human cells (Tooze, 1980). (2) A long segment of exogenous DNA can be inserted because the adenovirus genome is a double-stranded, linear DNA of about 36 kilobase-pairs (kb) and the information necessary for viral DNA replication and packaging locates at both termini of the genome. (3) Host protein synthesis is suppressed in a late stage of infection. Therefore, it is easy to detect and purify virus-coded products. (4) HeLa cells, one of the host cells of human adenovirus, can be grown in a spinner culture making it easy to work with on a preparative scale. (5) A major late promoter of adenovirus, which locates at 16.5 map units of the genome, is thought to be one of the strongest promoters in eukaryotes. More than 90 % of the synthesized protein in the late stage is coded for by the virus genome, and almost all the late mRNA is under the control of the major late promoter. Hexon, penton and fiber, which are major coat proteins of the adenovirus virions, are coded for by a late region of adenovirus and synthesized in prodigious quantities at a late stage of the normal lytic cell cycle after adenovirus infection. The amount exceeds that used for virus assembly and most remain in the nucleus as a multimeric form, which is observed as intranuclear paracrystals (Morgan et al. 1957). Such high expression is partly accounted for by the increased copy number of adenovirus genomes which are accumulated in infected nuclei by DNA replication. The tripartite leader sequence which is common in all the late mRNA is essential for an efficient translation (Thummel et al., 1983; Logan and Shenk, 1984). Therefore, in order to obtain the maximal expression, an exogenous gene should be positioned downstream of the major late promoter as well as the tripartite leader. Since there are no unique endonuclease cleavage sites at the preferable site because of the large genome size, we developed a special method to insert a nonselected exogenous gene into the desirable site of the adenovirus genome. Our method of constructing hybrid adenovirus had been developed using SV40 T antigen as a model exogenous gene because of its helper function (see below) (Thummel et al., 1981: 1982: 1983), and it was then improved so as to apply it to nonselected exogenous genes (Yamada et al., 1985: Yamada and Grodzicker, 1986).

III. CONSTRUCTION OF HYBRID ADENOVIRUSES

In order to insert an exogenous gene into a predetermined position in the adenovirus genome, a combination of in vitro and in vivo DNA recombination is used. Hybrid adenoviruses expressing SV40 T antigen can be selected for by their growth in monkey cells. This is because T antigen provides a helper function which allows human adenovirus to grow in normally nonpermissive monkey cells. If the foreign DNA to be inserted into the adenovirus genome is previously linked to a functional SV40 T-antigen gene, the hybrid virus carrying the foreign DNA can be selected for by its growth in monkey cells.

Using these features we constructed hybrid adenoviruses which carry exogenous genes in the middle of the third segment of the tripartite leader of the adenovirus genome as shown in Fig. 1. Starting material was a plasmid in which the coding sequence of the exogenous gene to be inserted was sandwiched between a piece of adenoviral genome and SV40 early DNA. The adenoviral segment contains a part of the third segment of the tripartite leader and corresponds to the sequences from BalI at 21.5 map units to XhoI at 26.5 map units. It serves as an element necessary for in vivo recombination to assure the junction between the exogenous gene to the third segment of the tripartite leader in the resulting hybrid adenoviruses. The SV40 early region

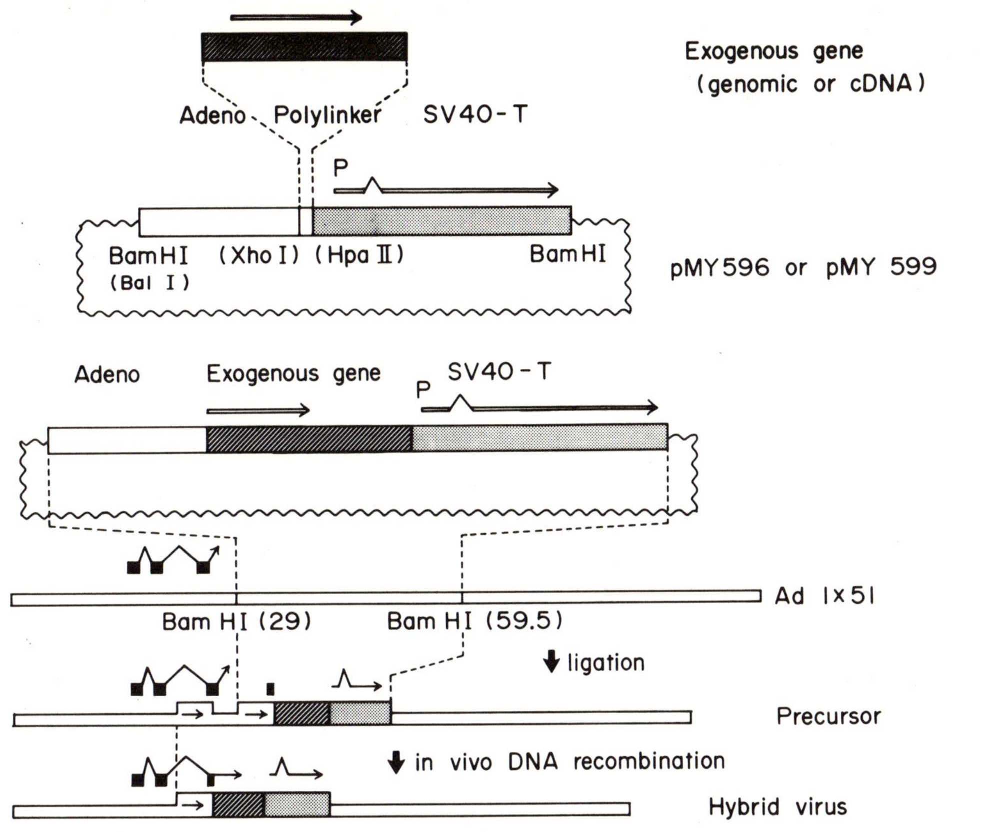

FIGURE 1. Construction of hybrid adenoviruses which carry the exogenous gene.

contains not only the entire T-antigen coding sequence but also the early promoter and replication origin. It serves as an element coding for the helper function. Expression of the SV40 T antigen is under the control of its own promoter in the resulting hybrid adenoviruses. Plasmid construction is facilitated by using pMY596 or pMY599 in which polylinker sequences (HindIII-XhoI-ClaI-EcoRI in pMY596 or HindIII-XbaI-ClaI-EcoRI in pMY599) link the adenoviral DNA and the SV40 T antigen gene. The entire adenovirus-exogenous gene-SV40 DNA fragment in the plasmid thus constructed was excised by BamHI digestion and ligated to BamHI-digested adenovirus lx51 DNA which has two BamHI sites at 29 and 59.5 map units. The ligation products were used to transfect cultured human cells along with lx51 helper DNA. Hybrid viruses expressing SV40 T antigen were selected on CV-1 monkey cells. After plaque purification on CV-1 cells, the genome structure of several candidate viruses was analyzed by restriction enzyme digestion and Southern blot hybridization. The hybrid virus carrying the exogenous gene in a correct configuration was then established.

The hybrid virus was defective because it lacked a part of the genome essential for virus growth. Thus a stock of the hybrid virus contained a mixture of hybrid and helper viruses. The proportion of the hybrid virus in a lysate was measured by comparing the number of plaque that appeared on human and monkey cells. The original lysate of hybrid adenoviruses (AdTkSVR591 and AdhCGSVR615, see below) contained only a small proportion (about 1 %) of the hybrid virus. After further plaque purification and growth of the hybrid virus, we obtained a virus lysate which contained a high proportion (15-30 %) of the hybrid virus.

IV. EXPRESSION OF GENE PRODUCTS USING THE ADENOVIRUS VECTOR SYSTEM.

Several hybrid adenoviruses carrying exogenous genes were obtained using the method described above. The thymidine kinase (Tk) gene of herpes simplex virus was chosen as the exogenous gene because its enzymatic activity can be measured easily. The alpha-subunit of human chorionic gonadotropin (hCG) was chosen as another exogenous gene. hCG which is a secreted glycoprotein hormone, is normally expressed in the pituitary and placenta, and ectopically expressed in some human tumor cells including HeLa cells. Therefore, we could compare the hCG product exogenously expressed by the adenovirus vector with endogenous hCG, when HeLa cells were used as host cells. A list of the hybrid adenoviruses we constructed is shown in TABLE I.

First we analyzed the expression of Tk by the hybrid adenovirus carrying the Tk gene (Yamada et al., 1985). When human or monkey cells infected with AdTkSVR591 were labeled with [^{35}S]-methionine and the cytoplasmic fraction was analyzed by immunoprecipitation, a protein of an apparent molecular weight of 47,000 dalton was detected at a late stage of infection. The same molecular weight species could be detected in herpes simplex virus infected cells at an early stage of infection. The level of Tk expression was high enough to be detected without immunoprecipitation. The Tk band detected in an SDS polyacrylamide gel was the third most abundantly labeled protein after the major capsid proteins, hexon and fiber, and could be detected even by staining with a dye. The Tk protein represented 10% of the newly synthesized protein in late infected cells and accumulated to represent 1% of total cellular protein under optimal conditions.

The Tk enzymatic activity in cell extract was assayed by measuring the extent of conversion of [^{3}H]-thymidine to [^{3}H]-dTMP. Levels of Tk activity in cells infected with AdTkSVR591 at a late stage were several times higher than in cells infected with herpes simplex virus at an early stage.

TABLE I. List of Hybrid Adenoviruses Constructed.

Hybrid viruses	Exogenous gene	Size of the exogenous DNA[a]
AdTkSVR591	Tk coding sequence[b]	2.4 kb
AdhCGSVR615	hCG genomic sequence	4.5 kb[c]
AdhCGSVR617	hCG genomic sequence	2.7 kb
AdhCGSVR619	hCG genomic sequence.	1.5 kb
AdhCGSVR641	cDNA of the hCG	0.7 kb
AdSVR599	no exogenous gene but carries the early region of SV40	

a. The hybrid viruses lost a portion of adenoviral DNA (26.5-59.5 map units, 11.9 kb) and gained the insert indicated above as well as 3.1 kb SV40 DNA.
b. The coding sequence is not interrupted by an intron.
c. The length of the hCG insert varied because a part of the intron and 3' noncoding region was deleted.

This result indicates that the Tk protein produced by the hybrid adenovirus had an enzymatic activity.

Adenovirus inhibits host protein synthesis at a late stage of infection. Various proteins in host cells are involved in the process of posttranscriptional and posttranslational modifications. To see whether such a modification system of the host cells was still active in a late stage when the maximal expression of the exogenous gene took place, hCG was expressed by infection of the hybrid adenovirus carrying hCG (Yamada and Grodzicker, 1986). HeLa cells produce low endogenous levels of hCG which is secreted into the culture medium. Both the cytoplasmic and the secreted hCG species (apparent molecular weight of 20,000, 18,000 and 23,000 dalton and 23,000 dalton, respectively) could be detected by immunoprecipitation with hCG antisera. When HeLa cells were infected with the hCG hybrid adenoviruses, the amount of hCG in both cell extracts and culture medium increased at a late stage of infection. There were no qualitative differences between virus-coded and endogenous hCG. All of the hCG species, whether virus-coded or endogenous, or whether detectable cytoplasmic species or secreted species, were proved to be glycosylated by labeling experiments in the presence or absence of tunicamycin and also by treatment of the product with endoglycosidase H. hCG was expressed by the hybrid virus not only in HeLa cells which normally produce hCG at low levels, but also expressed in other cells that do not express endogenous hCG. The hCG products in such host cells were glycosylated and secreted into the culture medium as was observed in HeLa cells.

Various hybrid adenoviruses listed in TABLE I, which carried different hCG segments, produced different levels of hCG. The highest hCG producer was AdhCGSVA615 and had hCG genomic DNA including 3' flanking sequences. The cytoplasmic hCG produced in AdhCGSVR615 infected cells represented about 0.8 % of newly synthesized cytoplasmic protein. This level was equivalent to that found for adenoviral major late capsid proteins and also that of Tk expressed by AdTkSVR591 after taking the number of methionine residues in a molecule into consideration . These results indicate that the gene products produced by this adenovirus vector system are correctly modified and processed.

V. CONCLUSION.

This adenovirus vector system is quite useful to produce protein at high levels that are biologically active, and correctly modified. Mansour et al. (1985) applied this system to produce small, middle and large T antigens of polyoma virus. The level was high enough to be used for an origin binding assay but less than that obtained by hybrid adenoviruses carrying Tk and hCG, because a lysate containing a small proportion of the hybrid virus was used. However, adenovirus is known to transform cultured rodent cells and cause tumors in animals. Most expression systems in mammalian cells that have so far been developed rely on tumor viruses because tumor viruses have been well studied. Virus vector systems should be further improved before they are used for production of materials for human use.

ACKNOWLEDGMENTS

Most of the work for developing the adenovirus vector was done at Cold Spring Harbor Laboratory. I thank Drs. J. D. Watson and T. Grodzicker for their useful discussions and encouragement.

REFERENCES

Elder, J.T., Spritz, R.A., and Weissman, S.M. (1981) Annu. Rev. Genet. 15:295.

Friedmann, T. (1983) "Gene Therapy: Fact and Fiction in Biology's New Approaches to Disease." Cold Spring Harbor Laboratory, New York.

Gluzman, Y. (1982) "Eukaryotic Viral Vectors." Cold Spring Harbor Laboratory, New York.

Harris, T.J.R. (1983) in "Genetic Engineering" (R. Williamson ed.) Vol. 4. p.127. Academic Press, London.

Logan, J. and Shenk, T. (1984) Proc. Natl. Acad. Sci. USA 81:3655.

Mansour, S.L., Grodzicker, T. and Tjian, R. (1985) Proc. Natl. Acad. Sci. USA 82:1359.

Morgan, C., Goodman, G.C., Rose, H.M., Howe, C. and Huang, J.S. (1957) J. Biophys. Biochem. Cytol. 3:505.

Ribgy, P.W.J. (1982) in "Genetic Engineering" (R. Williamson ed.) Vol. 3. p.83. Academic Press, London.

Thummel, C., Tjian, R. and Grodzicker, T. (1981) Cell 23:825.
Thummel, C., Tjian, R. and Grodzicker, T. (1982) J. Mol. Appl. Genet. 1:435.
Thummel, C., Tjian, R., Hu, S.-L. and Grodzicker, T. (1983) Cell 33:455.
Tooze, J. (1980) "DNA Tumor Viruses; Molecular Biology of Tumor Viruses." Cold Spring Harbor Laboratory, New York.
Yamada, M., Lewis, J.A. and Grodzicker, T. (1985) Proc. Natl. Acad. Sci. USA 82:3567.
Yamada, M. and Grodzicker, T. (1986) Mol. Cell. Biol. (in press).

MOLECULAR ANALYSIS OF THALASSEMIAS

Yasuyuki Fukumaki, Yoshihiro Takihara, Takanori Nakamura

Department of Biochemistry
Kyushu University School of Medicine, Fukuoka

Hideo Yamada

Department of Internal Medicine
Nagoya University School of Medicine, Nagoya

Yoshiro Ohta

Saga Koseikan Hospital, Saga

Shigeru Fujita

Department of Internal Medicine
Ehime University School of Medicine, Ehime

Yasuyuki Takagi

Laboratory of Molecular Genetics
Fujita Gakuen Health University School of Medicine, Toyoake

INTRODUCTION

Thalassemia is a heterogenous hereditary disease characterized by an impairment of the synthesis of one, or in some forms more than one of the globin polypeptide chains. Elucidation of the nature of the molecular defects underlying these conditions provides insights into various aspects of eukaryotic gene expression as well as basic data for establishing prenatal DNA diagnosis.

There are some approaches to the study of the molecular defects in thalassemias. One is structural analysis of the globin gene by Southern blotting, molecular cloning and DNA sequencing. However, structural analysis alone sometimes clearly cannot explain the relation between the structural change and the phenotype. Therefore functional analysis of the gene is indispensable in some cases. The study of expression of the gene in erythroid tissues derived from thalassemia patients is useful for this purpose. However, this approach is not necessarily useful because of heterozygosity of patients for different mutations and difficulties in obtaining suitable and sufficient blood or marrow samples from patients. Functional analysis of the gene by introducing the cloned genes into cultured heterologous cells is a practical alternative to the direct study of the genes in patients. This approach has a number of features. First, the globin gene is normally expressed only in erythroid cells. However, when the globin gene is cloned into the SV40 vector containing the tandemly repeated 72bp sequence, it is transcribed specifically and efficiently. Thus relative large amounts of globin RNA can easily be obtained. Second, it is possible to formally prove that abnormal expression is due to a sequence change within the cloned gene rather than at a distant location. Third, the expression of the mutant gene can be studied in the absence of its normal counterpart.

In this report, we will present the use of the heterologous expression system in the study of β-and δ -thalassemias found in Japan and discuss the validity and limitations of this approach in analyzing thalassemias.

MATERIALS AND METHODS

Patients. Patient 1 was a Japanese with homozygous β^{+}-thalassemia (reduced synthesis of the normal β-globin chain). Patients 2, 3 and 4 were independent Japanese patients with homozygous δ^{0}-thalassemia (absent synthesis of the normal δ -globin chain) and Patient 5, who was a sister of Patient 4, was a δ -thalassemia trait. Hematologic data of each patient were shown in Table 1. Patients 2, 3, 4 and 5 correspond to II-4 of family M, I-1 of family Iz, II-4 of family O and II-1 of family O described in ref. 1, respectively. The ratio of β- to α-globin chain synthesis in the peripheral erythroid cells of

Table 1. Hematologic data of each patient

	Hb (g/dl)	HbA_2 (%)	HbF (%)	MCV (um^3)	MCH (pg)	MCHC (g/dl)
Patient 1	7.2	5.9	14.0	74.4	16.7	22.5
Patient 2	13.0	0	4.7	91.7	29.1	31.7
Patient 3	10.7	0	0.9	92.8	32.0	34.5
Patient 4	13.6	0	1.0	85.2	28.3	33.2
Patient 5	12.6	1.4	0.9	80.8	29.1	36.0

Patient 1 is homozygous β^+-thalassemia. Patients 2, 3 and 4 are homozygous δ^0-thalassemia. Patient 5 is heterozygous δ-thalassemia.

Patient 1 was 0.34.

<u>DNA polymorphism haplotype analysis</u>. High molecular weight DNA was prepared from each patient as described (2). Restriction site polymorphisms in the β-globin gene cluster were assessed by Southern blot method using appropriate probes that have been described previously (3).

<u>Gene cloning and DNA sequencing</u>. The β-globin gene of Patient 1 was cloned in phage Charon 28 as a 7.8kb <u>HindIII</u> fragment that includes the structural gene and immediate flanking regions. The δ-globin gene from Patients 2, 3 and 4 were cloned in phage Charon 4A as DNA fragments partially digested with <u>EcoRI</u> that contain the δ- and β-globin genes. The nucleotide sequence of the β-globin gene of Patient 1 was determined by the method of Maxam-Gilbert for end-labeled DNA fragment (4). The δ-globin genes of Patient 2, 3 and 4 were subcloned in pUC9 and sequenced by the dideoxy chain termination method (5).

<u>Transient expression in COS cells</u>. Plasmid vector, pSVori 2172 (6) carrying the SV40 <u>PvuII-BglII</u> fragment that contains the origin of DNA replication and the tandemly repeated 72 bp sequence was used as a vector for expression of the globin gene in COS cells. The 4.9kb <u>BglII</u> fragment containing the β-globin gene and the 2.3kb <u>PstI</u> fragment

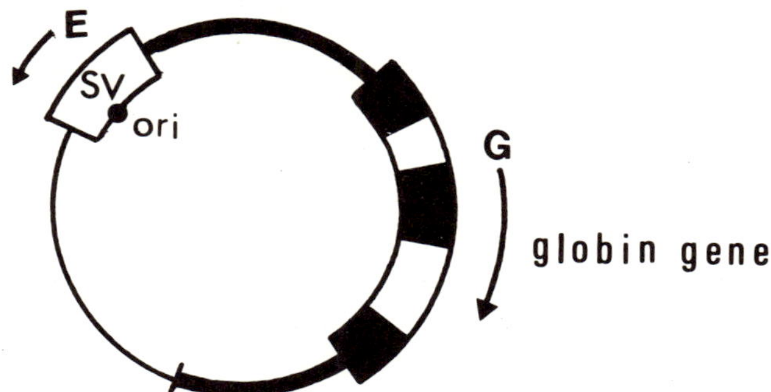

Fig. 1. Structure of the SV40 recombinant carrying the globin gene. SV: the SV40 PvuII-BglII fragment containing tandemly repeated 72-bp sequence, ori: replication origin of the SV40 DNA. Arrows E and G indicate polarity of SV40 early transcription unit and globin gene transcription unit, respectively.

containing the δ-globin gene were cloned at the BamHI site of pSVori 2172 as shown in Fig. 1. COS cells (7) were grown in Dulbecco's modified Eagle's medium with 10% newborn calf serum. Twenty micrograms of the β- or δ-globin gene recombinant and 2 μg of an α-globin gene recombinant (8) as an internal reference were added as calcium phosphate precipitates (9) to each 10-cm culture dish of subconfluent COS cells. Total cellular RNA was prepared from the transfected cells after incubation for 48 hours using the guanidinium method (10). Probe excess-S1 nuclease mapping of the total cellular RNA was performed using appropriate probes labeled at their 5' termini (11).

RESULTS

1) Analysis of the β^+-thalassemia

DNA polymorphism haplotype of the β-globin gene cluster. High molecular weight DNA isolated from Patient 1 was analyzed with respect to common DNA polymorphisms within the β-globin gene cluster. Patient 1 was homozygous at seven different polymorphic restriction sites (12) as shown in Fig. 2. Homozygosity for one haplotype and the consanguineous marriage of her parents strongly suggested that the patient was homozygous for a particular type of the

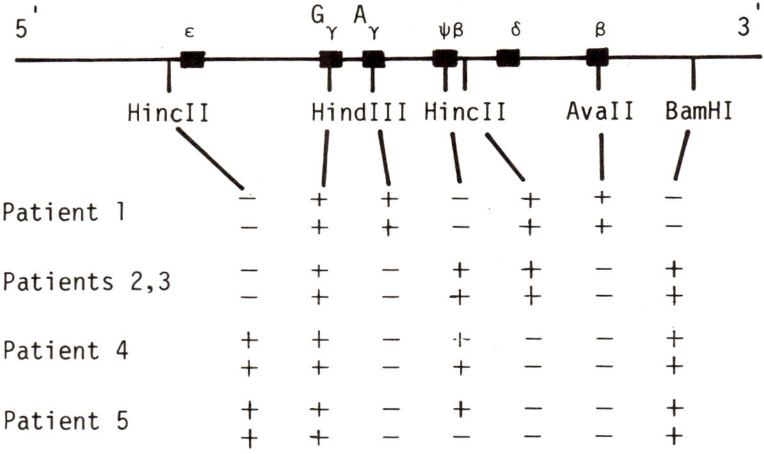

Fig. 2. Haplotype of the β-globin gene cluster of a patient with β^{+}-thalassemia (Patient 1) and patients with δ^{0}-thalassemia (Patients 2-5). + and - indicate the presence and absence of a restriction site, respectively.

β-globin gene mutation. Thus we chose a single cloned β-globin gene for further analysis.

DNA sequence of the β-globin gene region. The DNA sequence of the β-globin gene region from 200 bases 5' to the cap site to 200 bases 3' to the polyadenylation site was determined. The only notable feature was the presence of a single base substitution (A-G) at position -31, when compared with the known sequence (13) (Fig. 3). One additional substitution (G-T) was observed at position 74 in IVS2. This substitution has been reported to be due to a DNA polymorphism (12).

Expression of the β-thalassemia gene in COS cells. The A-G substitution is located within the TATA box (CATAAAA to CGTAAAA) which is known as a conserved sequence for efficient and accurate transcription of many eukaryotic protein-coding genes (14). Thus the A-G substitution could cause a thalassemia phenotype in the patient. To investigate this possibility we examined the expression of the gene in COS cells. Plasmid recombinants containing the normal and thalassemia β-globin genes were transfected in parallel into COS cells as described in the MATERIALS AND

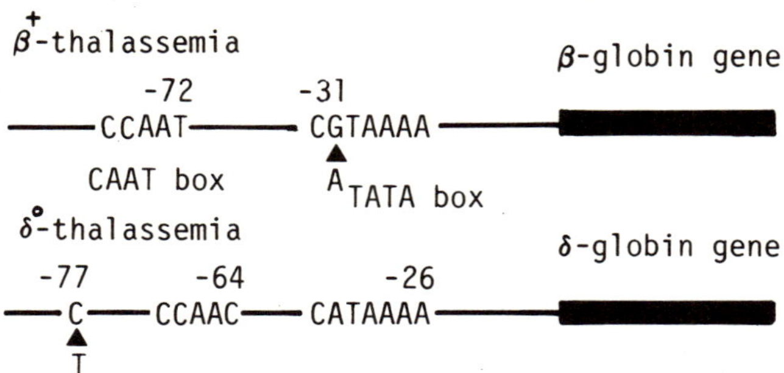

Fig. 3. Nucleotide sequence changes found in the 5' flanking region of the β-globin gene of β^{+}-thalassemia patient and the δ-globin genes of δ^{0}-thalassemia patients. Location of specific nucleotides is identified with respect to the cap site.

METHODS. Quantitative S1 nuclease mapping of total cellular RNA from the transfected COS cells showed a 45% reduction of β-globin RNA produced by the mutant gene relative to normal, when normalized for the difference in α-globin probe signal (Fig. 4). The S1 nuclease mapping with the 5' part of the β-globin probe demonstrated no differences between the mutant and the normal genes at the initiation site of transcription (data not shown).

2) Analysis of the δ^{0}-thalassemia

DNA polymorphism haplotype of the β-globin gene cluster. Southern blotting analysis of high molecular weight DNAs from Patients 2, 3 and 4 using the δ-globin probe showed normal restriction pattern, indicating no gross rearrangement or deletion of the δ-globin gene region (data not shown).

To examine homozygosity for the δ-globin allele, we characterized the DNA of three patients with respect to common DNA polymorphisms within the β-globin gene cluster (12). As shown in Fig. 2, Patients 2, 3 and 4 were homozygous at seven different polymorphic restriction sites. Patients 2 and 3 had the same haplotype that was different from the haplotype of Patient 4. A δ-thalassemia trait, Patient 5 had two different haplotypes, one of which was the same as that possessed by Patient 4. Homozygosity for a

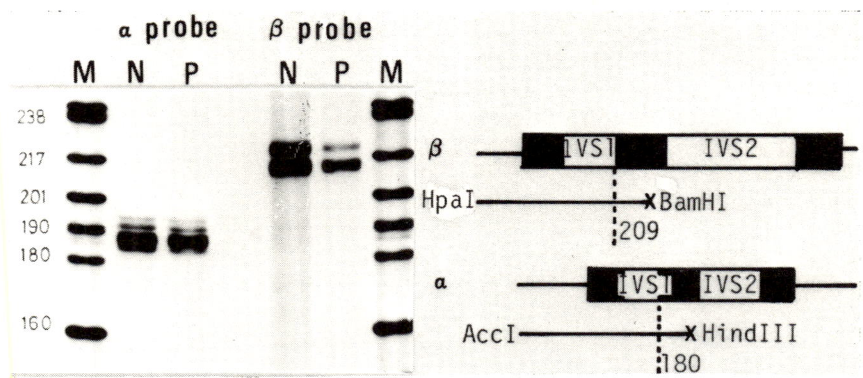

Fig. 4. Expression of the normal and β^{+}-thalassemia globin genes in transfected COS cells.

haplotype suggested homozygosity for a particular type of δ-globin gene mutation. Thus we chose a single δ-globin gene clone from each patient for further analysis.

<u>DNA sequence of the δ-globin gene region</u>. We sequenced the region from 449 bp upstream from the cap site to 207 bp 3' to the polyadenylation site of each δ-globin gene isolated from Patient 2, 3 and 4. There was no nucleotide sequence difference among these δ-globin genes. The nucleotide sequence of these genes differ from the published δ-globin sequence in four positions (15). They are a T-C change at position -77, a C-T change at position -199, an A addition in 4 bases strech of adenosine residue located from 408 to 411 and an AT dinucleotide insertion between positions -447 and -448. We determined the nucleotide sequence of the δ-globin genes isolated from two Japanese with no hematologic abnormalities. These nucleotide changes except for the substitution at position -77 were considered to be nucleotide sequence polymorphisms, since they were also present in the δ-globin genes of the normal subjects. Therefore a T-C substitution at position -77 is a common change among the δ-globin genes of three homozygous δ^{0}-thalassemia patients (Fig. 3).

<u>Expression of the δ^{0}-thalassemia gene in COS cells</u>. To investigate the correlation between the base substitution at -77 and the phenotype, SV40 recombinants containing the

δ -globin genes derived from either Patient 2 or the normal individual, were transfected into COS cells. Total cellular RNA was analyzed by S1 nuclease mapping from the viewpoint of transcriptional efficiency and processing of RNA. The S1 nuclease mapping using the DNA fragment covering the 5' side of the δ-globin gene as a probe demonstrated that fragments corresponding to the part of the exon 2 were protected and the level of δ-globin RNA transcribed from the patient globin gene was similar to that from the normal δ-globin gene (Fig. 5). Further S1 nuclease mapping (data not shown) showed no difference between expression of the patient and normal genes in the initiation site of transcription, the splice site of IVS2 and the cleavage site for polyadenylation.

DISCUSSION

Analysis of the β^+-thalassemia. In this report we presented a combination of structural and functional analyses to elucidate the molecular basis of two types of thalassemia, the β^+-and δ^0-thalassemias found in Japan. Structural analysis of the β^+-thalassemia gene showed a modified TATA box (ATAAAA-GTAAAA) of the β-globin gene. Functional analysis of this gene upon transient expression in COS cells revealed that this gene directed the production of 2-fold less β-globin RNA than the normal β-globin gene. These results indicate that the mild β^+-thalassemia phenotype of Patient 1 is the result of the TATA box mutation affecting RNA transcription.

Breathnach and Chambon (14) surveyed the TATA box sequence of 60 different eukaryotic genes and found that an adenine residue was present at the second position of the TATA box in 58 eukaryotic genes. In one case, a guanine residue was located at the corresponding position. However, results obtained from the analysis of the β^+-thalassemia gene presented here, established the functional significance of the second base of the TATA box for transcription of the human β-globin gene *in vivo*.

The extent to which the heterologous cell expression system reflects the phenotype caused by the mutation in normal erythroid cells is important in analysis of the mutant gene by using this system. In the case of the thalassemia studied here we had no opportunity to measure the level of β- and α-globin RNA in patient's erythroid

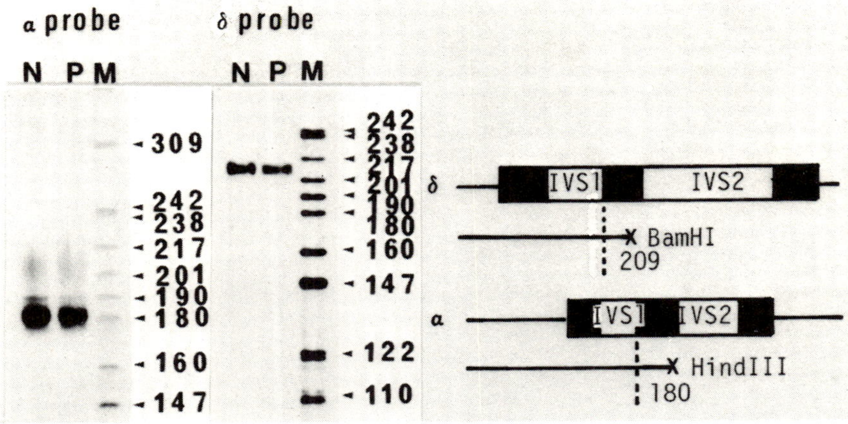

Fig. 5. Expression of the normal and the δ^0-thalassemia globin genes in transfected COS cells.

cells. However, the deficiency of β-globin RNA in heterologous cells seems to be comparable to that in erythroid cells because the ratio of β- to α-globin chain biosynthesis was 0.34 in peripheral erythroid cells of this homozygous patient. These speculations are supported by results of analyses of three different TATA box mutations carrying a β^+-thalassemia phenotype (16, 17, 18). Results presented here indicate that the heterologous cell expression system is valid for analysis of the TATA box mutations of thalassemias. Transient expression assay in heterologous cells have been demonstrated to be valid for analysis of promoter mutations in the -100 region of the β-globin gene (19, 20). RNA splicing mutations (20-22), RNA cleavage mutations (23) and premature termination mutations (24, 25) have also been represented by the transient expression system in heterologous cells. However, differences in content of abnormally processed RNA species were observed between heterologous cells and patients' erythroid cells (21-23, 26). These phenomena could be due to differences in stability of abnormal RNA between two kinds of cells mentioned above.

Analysis of the δ^0-thalassemia. In the structural analysis of the cloned δ-globin genes from homozygotes for δ^0-thalassemia, we found no apparent nucleotide changes within the coding region and IVS of the δ-globin gene, but a single base substitution (T-C) at position -77 in the 5' flanking region. When introduced into COS cells, the gene was expressed at normal levels with proper processing of RNA. These results indicate that the phenotype observed in three different patients is not due to a defective promoter, a defective RNA processing or a premature chain termination.

Recently two cases of thalassemia were reported in which the heterologous cell expression did not reflect the phenotype. In Dutch γβ-thalassemia (27) an extensive deletion in the ε-γ-δ globin gene region was detected. Although the β-globin gene of the patient has a normal sequence and functions properly upon transient expression in heterologous cells, the affected β-globin gene is present in an inactive configuration in the erythroid tissue. These observations were explained by either translocation of a normally inactive locus next to the β-globin gene in the affected chromosome or the deletion of the sequences which are normally required for the maintenance of the active state. In the case of δ^0-thalassemia studied here, there was no evidence of a major deletion or DNA arrangement in the β-globin gene cluster. Another case was the β-thalassemia silent carrier allele found in an Albanian family (28). Haplotype analysis of the Albanian family suggested that the silent carrier allele is not linked to the β-globin gene cluster. There might be abnormalities in trans-acting factors that affect the expression of the β-globin gene. Three patients with homozygous δ^0-thalassemia showed homozygosity for two kinds of haplotype. Since the region 10kb 5' to the β-globin gene is thought to be a relative hot spot for recombination in the β-globin gene cluster (29), a crossing over event in the region 5' to the β-globin gene probably accounts for the existence of two different haplotypes associated with the single base substitution at position -77 of the δ-globin gene. Furthermore, the δ-thalassemia trait who was the sister of Patient 4, inherited the haplotype possessed by Patient 4 and another haplotype that was not observed in three homozygous patients studied here. These results suggest that the δ-thalassemia locus is linked to the β-globin gene cluster, especially one base substitution at position -77 of the δ-globin gene. However, the possibility that the single substitution is just a polymorphism and the true locus is located in other region can not be eliminated.

The functional consequences of the substitution remain speculative. The involvement of sequences 5' to the γ-globin genes in hemoglobin switching was postulated based on a point mutations found at position -202 of the over expressed Gγ-globin gene in GγHPFH (Hereditary Persistence of Fetal Hemoglobin) (30) and positions -117 and -196 of the

Aγ-globin gene of Greek type HPFH (31, 32) and Southern Italian HPFH (33), respectively. Particularly, a base substitution at -117 of inappropriately expressed Aγ-globin gene in adult life suggests that the CAAT box or its surrounding sequence has a role in developmental control of γ-globin genes. Recently Emerson et al. (34) demonstrated that the trans-acting factor that was detected at proper stage and interacted with the region -40 to -200 bp in the 5' flanking region of the chicken adult β-globin gene. These results indicate that a DNA region -200 bp 5' to the globin gene is important in a cell -and developmental stage-specific expression of the globin gene. The single base substitution detected in three δ^0-thalassemia patients is located 9 bp 5' to the CAAT box of the δ-globin gene. The region around the CAAT box might be the binding site of a positive trans-acting factor for the δ-globin gene expression. One base substitution at position -77 may inhibit the DNA-protein interaction, hence turning off expression of the δ-globin gene. To investigate these possibilities the functional study of this δ-globin gene in erythroid cells under differential control of its expression is necessary. This type of analysis is now in progress.

In this report we presented the use of structural analysis and functional analysis of the cloned gene in heterologous cells in the study of thalassemias. Molecular defects that result in dysfunction of the proximal promoter element, the abnormal splicing, the abnormal cleavage of the globin RNA and premature termination of translation were elucidated by use of two types of approach mentioned above. However, there are some cases of thalassemia whose molecular defects remain to be unclear even after the structural and functional analyses of the cloned gene. Some of these molecular defects could be elucidated by the gene expression study in homologous cells. It is likely that results obtained by such study will add to our understanding of the regulation of the globin gene expression in vivo.

ACKNOWLEDGMENTS

We thank Dr. T. Maniatis for providing recombinant plasmids of the normal globin genes, Drs. K. N. Subramanian and D. H. Hamer for gifts of pSVori 2172 and the SV40- α-globin gene recombinant, respectively. We thank Ms. Honoho Hamada for expert preparation of the manuscript. This work was supported in part by the special project research grant, No. 60127010 (inborn errors of metabolism) from the Ministry of Education in Japan.

REFERENCES

1. Ohta, Y., Yasukawa, M., Saito, S., Fujita, S., and Kobayashi, Y. (1980). Hemoglobin. 4:417.
2. Kimura, A., Ohta, Y., Fukumaki, Y., and Takagi, Y. (1984). Biochem. Biophys. Res. Comm. 119:968.
3. Matsunaga, E., Kimura, A., Yamada, H., Fukumaki, Y., and Takagi, Y. (1985). Biochem. Biophys. Res. Comm. 126:185.
4. Maxam. A., and Gilbert, W. (1980). Meth. Enzymol. 65:499.
5. Hattori, M., and Sakaki, Y. (1986). Anal. Biochem. 152:232.
6. Bergsma, D. J., Olive, D. M., Hartzell, S. W. and Subramanian, K. N. (1982). Proc. Natl. Acad. Sci. USA. 79:381.
7. Gluzman, Y. (1981). Cell. 23:175.
8. Felber, B. K., Orkin, S. H. and Hamer, D. H. (1982). Cell. 29:895.
9. Wigler, M., Silverstein, S., Lee, L.S., Pellicer, A., Cheng, Y. C., and Axel, R. (1977). Cell. 11:223.
10. Glisin, V., Crkvenjakov, R., and Byus, C. (1974). Biochemistry. 13:2633.
11. Berk, A. J., and Sharp, P. A. (1977). Cell. 12:721.
12. Orkin, S. H., Kazazian, H. H. Jr., Antonarakis, S. E., Goff, S. C., Boehm, C. D., Sexton, J. P., Waber, P. G., Giardina, P. J. V. (1982). Nature. 296:627.
13. Lawn, R. M., Efstratiadis, A., O'Connell, C., Maniatis, T. (1980). Cell. 21:647.
14. Breathnach, R., and Chambon, P. (1981). Ann. Rev. Biochem. 50:349.
15. Poncz, M., Schwartz, E., Ballantine, M., and Surrey, S. (1983). J. Biol. Chem. 258:11599.
16. Surrey, S., Delgrosso, K., Malladi, P., and Schwartz, E. (1985). J. Biol. Chem. 260:6507.
17. Orkin, S. H., Sexton, J. P., Cheng, T -c., Goff, S. C., Giardina, P. J. V., Lee, J. I., and Kazazian, H. H. Jr. (1983). Nucl. Acids Res. 11:4727.
18. Antonarakis, S. E., Orkin, S. H., Cheng, T -c., Scott, A. F., Sexton, J. P., Trusko, S. P., Charache, S., and Kazazian, H. H. Jr. (1984). Proc. Natl. Acad. Sci. USA. 81:1154.
19. Orkin, S. H., Antonarakis, S. E., Kazazian, H. H. Jr.

(1984). J. Biol. Chem. 259:8679.
20. Treisman, R., Orkin, S. H., and Maniatis, T. (1983). Nature. 302:591.
21. Fukumaki, Y., Ghosh, P., Benz, E. J. Jr., Reddy, V. B., Lebowitz, P., Forget, B. G., and Weissman, S. M. (1982). Cell. 28:585.
22. Moschonas, N., Boer, E., Grosveld, F. G., Dahl, H. H. M., Wright, S., Shewmaker, C. K., and Flavell, R. A. (1981). Nucl. Acids Res. 9:4391.
23. Higgs, D. R., Goodbourn, S. E. Y., Lamb, J., Clegg, J. B., Weatherall, D. J., and Proudfoot, N. J. (1983). Nature. 306:398.
24. Takeshita, K., Forget, B. G., Scarpa, A., and Benz, E. J. Jr. (1984). Blood. 64:13.
25. Humphries, R. K., Ley, T. J., Anagnou, N. P., Baur, A. W., and Nienhuis, A. W. (1984). Blood. 64:23.
26. Maquat, L. E., and Kinniburgh, A. J. (1985). Nucl. Acids Res. 13:2855.
27. Kioussis, D., Vanin, E., deLange, T., Flavell, R. A., and Grosveld, F. G. (1983). Nature. 306:662.
28. Semenza, G. L., Delgrosso, K., Poncz, M., Malladi, P., Schwartz, E., and Surrey, S. (1984). Cell. 39:123.
29. Kazazian, H. H. Jr., Orkin, S. H., Markham, A. F., Chapman, C. R., Youssoufian, H., Waber, P. G. (1984). Nature. 310:152.
30. Collins, F. S., Stoeckert, C. J. Jr., Serjeant, G. R., Forget, B. G., and Weissman, S. M. (1984). Proc. Natl. Acad. Sci. USA. 81:4894.
31. Gelinas, R., Endlich, B., Pfeiffer, C., Yagi, M., and Stamatoyannopoulos, G. (1985). Nature. 313:323.
32. Collins, F. S., Metherall, J. E., Yamakawa, M., Pan, J., Weissman, S. M., and Forget, B. G. (1985). Nature. 313:325.
33. Giglioni, B., Casini, C., Mantovani, R., Merli, S.,Comi, P., Ottolenghi, S., Saglio, G., Camaschella, C., and Mazza, U. (1984). EMBO J. 3:2641.
34. Emerson, B. M., Lewis, C. D., Felsenfeld, G. (1985). Cell. 41:21.

IV. DNA DIAGNOSIS AND GENE THERAPY

MODEL STUDIES TOWARD HUMAN GENE THERAPY

T. Friedmann[1]

Department of Pediatrics
University of California, San Diego
La Jolla, California

Therapy currently available for most human genetic disease is largely ineffective. With the impressive recent technical and conceptual advances in human and molecular genetics, techniques have been developed during the past several decades that allow not only a detailed understanding of the nature of the genetic defect in many diseases but also the possible development of conceptually new approaches to therapy. One such approach, called gene therapy, is based on the presumption that the correction of a disease phenotype can be accomplished either by modification of the expression of a resident mutant gene or the introduction of new genetic information into defective cells or organs *in vivo*. At the present time, techniques for genetic correction or site-specific gene replacement are not yet well developed, and so most present models rely on the development of efficient gene transfer systems to introduce functional, normal, wild-type genetic information into genetically defective cells *in vitro* and *in vivo*. To be clinically useful, the availability of efficient and safe delivery vectors for foreign sequences must be augmented by easy accessibility of suitable target cells or organs and the development of techniques to introduce the vector stably into those target cells *in vivo*. My colleagues and I have used the gene for the purine biosynthetic enzyme hypoxanthine guanine phosphoribosyltransferase (HPRT) to determine if retroviral vectors can be used to transfer a functional gene into mutant human cells stably and efficiently,

[1]Supported by NIH grants GM28223 and HD20034 and by the Weingart and Keck Foundations.

if such a transduced and expressed gene can correct a disease phenotype *in vitro*, if the gene can then be transferred into suitable organs in whole animals, and if the new gene expression can correct a disease phenotype *in vivo*.

Purine biosynthesis in most mammalian cells (Figure 1) occurs by two major pathways (1). The de novo pathway begins with the coupling of glutamine and phosphoribosylpyrophosphate (PRPP) and leads, after a series of steps, to the production of the product at the hub of the biosynthetic pathway, inosinic acid. From there, guanylic and adenylic acids are produced and these are further used for nucleic acid synthesis. An alternative pathway, the reutilization or salvage pathway, begins with the preformed bases hypoxanthine and guanine and with phosphoribosylpyrophosphate, and in reactions catalyzed by HPRT, leads to the conversion of the bases to inosinic and guanylic acids. It's particularly interesting to note, too, that this same purine biosynthetic pathway also contain two of the other useful model disorders for human gene therapy, namely the enzymic deficiencies of adenosine deaminase which converts adenosine and deoxyadenosine to inosine and deoxyinosine, and of purine nucleoside phosphorylase which catalyzes the phorphorolysis of guanosine and inosine.

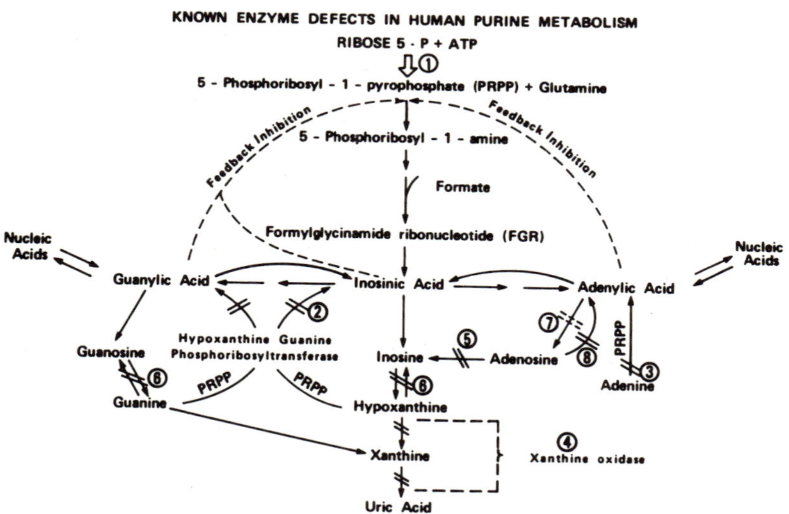

FIGURE 1. Pathway for purine synthesis in human cells. The de novo pathway and reutilization pathways lead to the synthesis of nucleotide precursors inosinic, adenylic and guanylic acids for subsequent production of nucleic acids.

Deficiencies of both of these enzymes lead to severe immunological defects (2).

One might expect that a deficiency of the enzyme HPRT would not lead to disease because of the availability of an alternative pathway for producing inosinic acid. However, as Dr. J. Seegmiller showed in 1967 (3), HPRT deficiency does lead to a severe human disease called Lesch-Nyhan Syndrome which is characterized by severe neurological symptoms, including choreoathetosis and varying degrees of mental retardation, hyperuricemia leading to renal destruction, and a peculiar and compulsive self-mutilation behavior in which affected boys bite the edges of their lips and amputate their fingertips by biting (4). The behavior is involuntary and painful, and can be managed only by removal of the teeth and the use of arm restraints. The metabolic damage caused by hyperuricemia can be treated reasonably effectively with allopurinol to reduce serum uric acid levels.

Some years ago, my colleagues and I chose to begin model studies with HPRT on gene transfer approaches to disease therapy because of the powerful selective systems available <u>in vitro</u> for cells that either express or fail to express the enzyme. We were able relatively quickly to clone the full length HPRT cDNA in the form of an expression vector developed by Hiroto Okayama and Paul Berg (5). When this purified plasmid vector was introduced into HPRT deficient mouse cells by calcium phosphate transfection, we were able to show that an occasional cell, approximately 1 in 100,000 recipient cells, had become resistant to aminopterin and therefore was producing HPRT enzyme activity. Furthermore we could demonstrate that the HPRT in these transfected mouse cells was human HPRT. Because efficiency was so low, this method of gene transfer would not be suitable for the extensive <u>in vivo</u> studies that we wished to undertake, and together with our colleagues Inder Verma and Dusty Miller, we chose to develop an efficient gene transfer system based on the use of retroviral vectors of the kind previously reported by Temin, Weinberg, Scolnick and their colleagues (6-8).

The acutely infectious murine retroviruses (9) consist of a single-strand RNA genome classically encoding three viral genes, namely the group-specific antigen (gag), RNA-dependant DNA polymerase (pol) and the envelope gene (env). After infection, the RNA genome is converted to a double-stranded DNA copy at both ends of which are identical regions of repeated sequence that have come to be called long terminal repeats (LTRs). The linear double-stranded DNA form is circularized into forms containing either one or two LTRs, and the double LTR molecule is then integrated into random

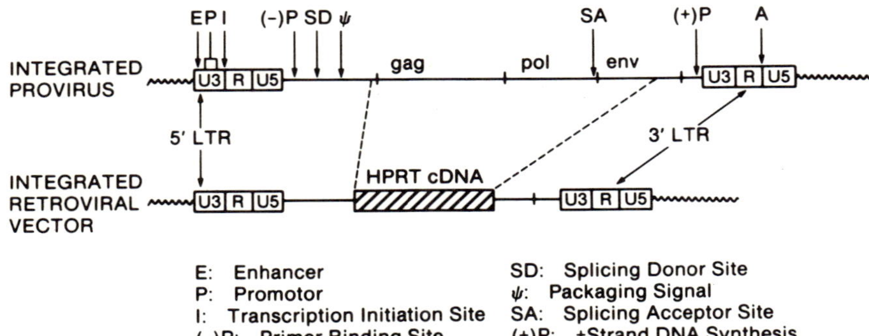

FIGURE 2. The structure and expression of the integrated provirus. The promoter and enhancer sequences within the 5' LTR, the nearby packaging and splice donor sites and initiation sites for DNA synthesis and the polyadenylation signal in the 3' LTR are retained in defective retroviral vectors.

sites of the host cell genome. The resulting integrated structure, called the provirus (Figure 2), is then expressed as new viral gene products from transcriptional signals including promoters and enhancers within the 5' LTR and transcriptional termination signals in the 3' LTR. Adjacent to the 5' LTR are other crucial sequences needed for virus replication and production, including the site for initiation of first strand of DNA replication, the sequence required for packaging of viral transcripts into virus particles, the so called ψ signal, and a sequence required for the initiation of second strand DNA synthesis near the 3' LTR. All of these sequences, plus the short sequence within the LTR that directs integration, must be preserved during the design of a integrating vector, but the viral genes gag, pol and env can be replaced by any full length, expressible cDNA or genomic sequence.

Such a recombinant molecule can be introduced into cells by calcium phosphate transfection to give approximately the same efficiency of gene transfer as with non-retroviral vectors. However, the naked retroviral vector DNA can be encapsidated into transmissible viral particles by several means, most notably by the use of packaging lines of cells (10,11) carrying a provirus expressing the viral gene products gag, pol and env but also containing a deletion

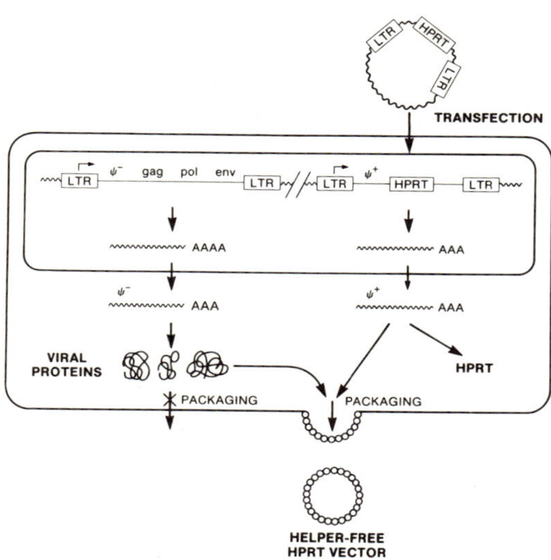

FIGURE 3. A cell carrying a provirus encoding the viral structural genes gag, pol and env but also having a deletion of the packaging signal ψ. The viral proteins therefore cannot package the proviral transcript in cis, but are able in trans to package transcripts derived from another provirus containing an ψ sequence.

in the ψ or packaging signal (Figure 3). Because of the absence of the packaging signal, transcripts from this constitutively expressed provirus cannot be packaged, but transcripts from another provirus can be packaged in trans by the viral gene proteins present in the cell to produce infectious transmissible virus containing not viral genes but the gene carried by the second provirus, in this case the HPRT gene. Transmissible HPRT virus produced in this way is able to infect a number of human and rodent cells in culture to express the HPRT gene, and as we showed some years ago in human lymphoblasts, to correct the metabolic abnormalities of HPRT-deficient human cells (12). The startling difference between the transmissible vector and the naked recombinant DNA molecule is the great increase in efficiency of the gene transfer event. Calcium phosphate transfection efficiency leads to genetic transformation of one cell per 10,000-100,000 recipient cells, while gene transfer with the transmissible viral vector achieves a gene transfer of approximately 100% of cells *in vitro*.

To begin to study the usefulness of HPRT as a model for in vivo gene replacement, we selected the mouse bone marrow as a suitable and useful target organ because of its easy accessibility and the presence of stem cells that serve to ensure long term expression of a genetic transformation. In this system, mouse bone marrow is exposed in vitro to high titer vector and then is returned to animals after total body lethal irradiation. Surviving mice must repopulate their marrow from infected cells in the culture, and therefore should express the foreign gene in their marrow and spleen cells and possibly peripheral blood cells as well. We were able to determine that the efficiency of infection of the bone marrow cells with the HPRT viral vector was quite high (13). However, while we were able to identify occasional samples that also in fact expressed the foreign human HPRT, the efficiency of gene expression was quite low. This severe problem of very low in vivo efficiency of gene expression from retroviral vectors is still characteristic of most mouse bone marrow repopulation systems using a number of different transduced genes. A great deal of work and effort is now being put into modification of the design of the vectors to try to enhance the expression in mouse and other bone marrow stem cells.

For the retroviral gene transfer model to be useful clinically, methods should be developed to introduce foreign genes not only into bone marrow but a variety of other organs as well. It is also important to demonstrate that the provirus integrated into the recipient cells is stable over a long period of time, and we have already demonstrated that at least the HPRT virus in human lymphoblasts does demonstrate a significant rate of mutation and reversion, up to approximately 4×10^{-5} per generation, and that a variety of genetic and epigenetic mechanisms are responsible for shutting down the HPRT gene expression in such cells (14). The frequency and mechanisms of shutdown of proviral expression by mutation or by epigenetic mechanisms are very similar to those previously described for the expression of the sarc gene in cells transformed by avian sarcoma viruses. While HPRT provirus loss or shutdown is certainly not excessively high, it is significantly greater than background rates.

An additional ideal goal for retroviral transduction is that the integrated provirus do no additional metabolic or genetic damage to the recipient cells. Since proviruses integrate at random in the host cell genome, it is inevitable that some cells will be damaged by insertional mutagenesis, either by the integration of the provirus into an essential gene to disrupt its function or the alteration of normal

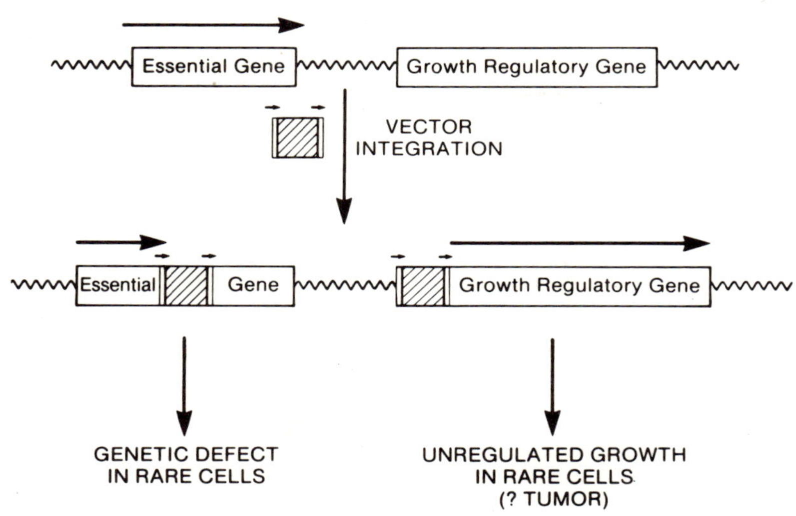

FIGURE 4. Insertional mutagenesis by integrated provirus, either by interruption of an essential resident gene or by promoter or enhancer insertion near a cellular gene.

regulation of nearby cellular genes caused by the introduction of proviral promoter or enhancer sequences (Figure 4). Until methods are perfected for carrying out site-specific integration, there seems to be no way to prevent the occasional disruption of cellular genes as a basis for insertional mutagenesis. However, modifications to the vector design may help to reduce cellular mutation and damage by promoter or enhancer insertion. We have approached this problem by deleting the promoter and enhancer regions of the 3' LTR by site-specific mutagenesis (Figure 5). After one round of replication, because the U3 region of the 3' LTR serves as a template for the synthesis of the corresponding region of the 5' LTR, deletions inserted into the U3 region of the 3' LTR are transferred to the 5' LTR (9). By combining these mutations with the introduction of an internal metallothionein promoter in front of the HPRT cDNA (Figure 6), we have been able to express the transduced human HPRT activity in a variety of cells from the internal promoter of a vector which is capable of integrating into the host cell genome but which has no enhancer or promoter activities within its own LTRs (15).

We are continuing to study the features of retroviral vector design which permit efficient expression and the

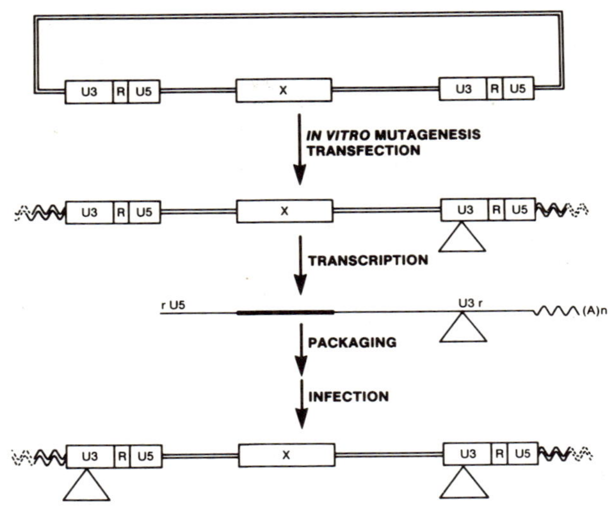

FIGURE 5. Scheme for the deletion of the transcriptional signals in the 5' LTR by site directed mutagenesis in the 3' LTR. The U3 region of the 3' LTR serves as template for the synthesis of the U3 region of the 5' LTR, and therefore deletions made in the 3' LTR will be expressed in the 5' LTR after retroviral replication.

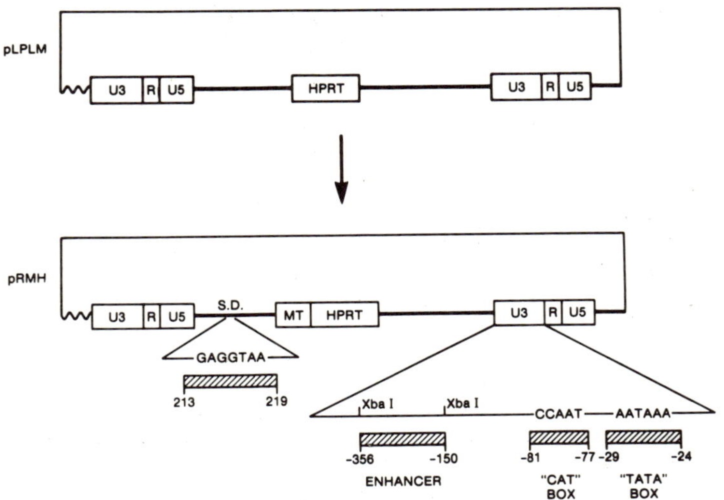

FIGURE 6. Target sequences for introducing deletions in the retroviral LTRs and the splice donor site and for the insertion of the metallothionein promoter to express the HPRT cDNA.

introduction of transduced genes into a variety of other recipient target cells. It seems very likely that in the relatively near future, we will know enough about the expression, stability, insertional mutagenic potential and other features of retroviral vectors to justify attempts in a limited number of clinical situations to provide genetic function to human beings suffering from genetic disorders (16-18). The long-term potential for a genetic attack on human disease is extremely great and exciting and limited only by our imaginations.

REFERENCES

1. Kelley, W., and Wyngaarden, J., In "Metabolic Basis of Inherited Disease." (J. Stanbury, J., J. Wyngaarden, D. Frederickson, J. Goldstein, and M. Brown, eds.), p. 1115-1143. McGraw Hill, New York, 1983.
2. Seegmiller, J., In "Physiological and Regulatory Functions of Adenosine and Adenine Nucleotides." (H. Baer and G. Drummond, eds.), p. 395-408. Raven Press, New York, 1977.
3. Seegmiller, J., Rosenbloom, F., and Kelley, W., Science 155:1682-1685 (1967).
4. Nyhan, W., Ann. Rev. Med. 24:41-56 (1973).
5. Jolly, D., Okayama, H., Berg, P., Esty, A., Filpula, D., Bohlen, P., Johnson, G., Shively, J., Hunkapillar, T., and Friedmann, T., Natl. Acad. Sci. USA 80:477-481 (1983).
6. Shimotohno, K., and Temin, H., Cell 26:67-77 (1981).
7. Wei, C., Gibson, M., Spear, P., and Scolnick, E., J. Virol. 39:935-944 (1981).
8. Tabin, C., Hoffman, J., Goff, S., and Weinberg, R., Mol. Cell Biol. 2:426-436 (1982).
9. "RNA Tumor Viruses." (R. Weiss, N. Teich, H. Varmus, and J. Coffin, eds.) Cold Spring Harbor Laboratory, New York, 1982.
10. Mann, R., Mulligan, R., and Baltimore, D., Cell 3:153-159 (1983).
11. Miller, A., Law, M.-F., and Verma, I., Mol. Cell. Biol. 5:431-437 (1985).
12. Willis, R., Jolly, D., Miller, A., Plent, M., Esty, A., Anderson, P., Chang, H., Jones, O., Friedmann, T., and Seegmiller, J., J. Biol. Chem. 259: 7842-7849 (1984).
13. Miller, A., Jolly, D., Friedmann, T., and Verma, I., Science 225:630-632 (1984).

14. Jolly, D., Willis, R., and Friedmann, T., Molec. Cell. Biol. 6:1141-1147 (1986).
15. Yee, J.-K., Jolly, D., Moores, J., Respess, J., and Friedmann, T., "Gene Expression From a Transcriptionally Disabled Retroviral Vector." Symposium of Quantitative Biology, v. XXX, Cold Spring Harbor Laboratory, New York, in press.
16. Friedmann, T., and Roblin, R., Science 175:949-955 (1972).
17. Friedmann, T., "Gene Therapy--Fact and Fiction in Biology's New Approaches to Disease?" Cold Spring Harbor Laboratory, New York, 1983.
18. Anderson, W., Science 226:401-409 (1984).

MOLECULAR MECHANISM OF THE FEMALE LESCH-NYHAN SYNDROME

Nobuaki Ogasawara[1]

Department of Biochemistry
Institute for Developmental Research
Aichi Prefectural Colony
Aichi 480-03, Japan

I. INTRODUCTION

A deficiency of the enzyme hypoxanthine-guanine phosphoribosyltransferase (HPRT; EC2.4.2.8) is associated with two clinical syndrome in human. A virtually complete deficiency of the enzyme occurs in patients with Lesch-Nyhan syndrome (1,2) and a partial deficiency of HPRT in severe form of gout (3). HPRT is X-linked and because of inability of reproduction in Lesch-Nyhan patient this syndrome occurs only in males. We have, however, an unusual case of a female Lesch-Nyhan patient, whose mother is not heterozygous for a deficiency of HPRT (4,5)

II. CASE REPORT

The patient was born in February, 1975. Nothing was particular during pregnancy and also at birth. Only special was that both parents were quite old when she was born; the father was 47 years old and the mother 33 years old. There were some neurological abnormalities until 5 years old, but Lesch-Nyhan syndrome was not doubt because the patient was a female. At 5 years old, she begun to bite her lips and fingers and lips were partially lost. Thus, Lesch-Nyhan syndrome was suggested and HPRT activity in erythrocytes was determined. As shown in Table I, the activity was not

[1]Supported by the grant from the Gout Research Foundation and Intractable Diseases Grant from the Ministry of Health and Welfare of Japan.

detectable. APRT activity was increased 2-3 folds compared to control as observed typically in male Lesch-Nyhan patient. HPRT activities of fibroblast and EB virus transformed B-lymphoblast from the patient were also undetectable.

III. RESULTS AND DISCUSSION

A. The Patient Is Female and Her Mother Is Not a Heterozygote of HPRT$^-$

To study on the genetic and molecular mechanism of this female patient, there are two important points to be confirmed. That is, 1) the patient is a real female, and 2) the mother is not heterozygous. The patient's karyotype was 46, XX with a rate replicating X, indicating that the patient is a female. The mother is not heterozygote (4,5), confirmed by 1) no HPRT negative hair follicle, 2) normal HPRT activity in fibroblasts, while one half of the activity in the known heterozygote, 3) no 6-thioguanine resistant cell in her fibroblasts and T-lymphocytes, and conclusively, 4) existence of two copies of HPRT genes in the mother but HPRT gene deletion in patient's maternal X chromosome.

B. Possible Mechanism for Female Lesch-Nyhan Patient, Whose Mother Is not Heterozygote

Fig. 1 shows three possible mechanisms. In A, her HPRT deficiency represents a mutation in a regulatory gene or a gene which would be complimentary to structural HPRT gene. The existence of this kind of gene necessary for the expression of the HPRT locus is suggested (6-9). If the girl is deficient in a regulatory gene, fusion of her cells with the cells from the male HPRT deficient patient would furnish the regulatory gene and the fused heterokaryons would produce active enzyme. These experiments, however, showed that there was no complimentary nature between the female and the male

TABLE I. HPRT and APRT Activities in Erythrocytes

	HPRT	APRT
	nmoles/min/mg Hb	
Normal (9)	1.73 ± 0.26	0.323 ± 0.075
Patient	<0.003	0.809
Mother	1.19	0.272
Father	1.58	0.227

patients with Lesch-Nyhan syndrome. Also, if the mechanism is correct, fusion of the patient cells with LTK, a thymidine kinase less derivative of mouse cell, would produce human HPRT. Result was also negative.

B and C show other two mechanisms. B is the double mutation mechanism. In this mechanism, we need several hundred years to obtain one female Lesch-Nyhan patient, assuming 10^8 birth/year in the world. In C, the assumption is a HPRT gene mutation on one of the X chromosome and the specific inactivation of the other X chromosome which presumely carries a functional gene. This is the one against Lyon's random inactivation theory.

C. The Studies on mRNA and DNA

Northern analysis was carried out and showed that this patient was $mRNA^-$ phenotype.

Southern analysis after Pst I digestion is useful for analysis of HPRT gene mutation, since the corresponding exon was known for each of band of Southern blot (10). Pst I digestion showed that the patient has a complete HPRT gene, but interestingly the densities of each band were likely to

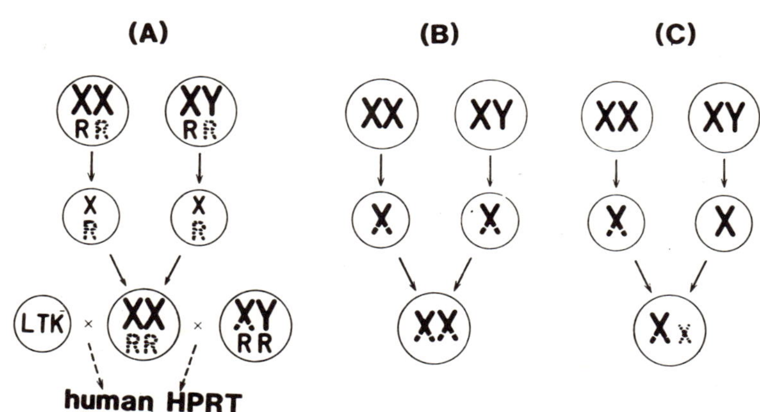

FIGURE 1. Possible mechanism of a female Lesch-Nyhan patient.

those of the father with one X chromosome, rather than the mother with two X chromosomes. These results indicate the total HPRT gene deletion on one of the X chromosome and the minor or undetectable change by Southern on the other X chromosome. In Bam HI digestion, restriction length polymorphism was described (11). These are 22 Kb/ 25 Kb pair, 22 Kb/ 18 Kb pair and 12 Kb/ 25 Kb pair. Therefore, under some combinations it is possible to identify the maternal or paternal origin of the total HPRT gene deleted X chromosome. The patient and her both parents showed the same rare 12 Kb/ 25 Kb pair. Thus, it was not possible to identify the HPRT gene deleted X chromosome.

Boggs and Nussbaum (12) described the restriction length polymorphism of anonymous human X-chromosome specific gene DXS10 after digestion by Taq I. One is 5 Kb/ 3.5 Kb pair type and the other 7 Kb/ 3.5 Kb pair. Southern blot showed that the mother was 7 Kb/ 3.5 Kb homozygous, the father 5 Kb/ 3.5 Kb hemizygous and the patient 7 Kb/ 5 Kb/ 3.5 Kb heterozygous. These results indicate that each of two X chromosomes of the patient orginates from the mother and from the father, and the analysis of DXS10 polymorphism promises to identify the origin of X chromosomes.

Thus, to find out on which of the X chromosomes the HPRT gene is totally deleted, the patient fibroblasts were fused with LTK and the cells were selected in HAT medium. DXS10 analysis showed that of 111 hybrid clones, 2 clones had both maternal and paternal X chromosomes and 4 had the paternal X chromosome alone. By recloning of two clones having both X chromosomes, one clone having the maternal X chromosome alone and two additional clones having the paternal X chromosome alone were obtained (Table II). When these clones were analyzed for HPRT gene after digestion with Pst I, the clones having the paternal X chromosome had human HPRT gene, but the clone having the maternal X chromosome alone deleted totally human HPRT gene.

These results showed the following possibilities; 1) this patient may represent the rare case of two independent mutations on both X chromosome (Fig. 1b)-one total HPRT gene deletion on the maternal X chromosome and a point mutation or a minor deletion on the paternal X chromosome which is not detectable in Southern blot yet results in a non-functional gene, since Northern blot analysis showed no mRNA in the patient lymphoblast; 2) this patient's disease may result from a total HPRT gene deletion on the maternal X chromosome and a selective inactivation of the paternal X chromosome that would presumably carry functional HPRT gene (Fig. 1c). To test these two possibilities, glucose-6-phosphate dehydrogenase (G6PDH) of the clones having human X chromosome(s) was analyzed (Table I). The clones with two human X chromosomes expressed human G6PDH. One clone having the maternal X chromosome alone also expressed the human enzyme. Six clones having the paternal X chromosome alone did not express human G6PDH. These results indicate the existence of only inactive paternal X chromosomes and a selective inactivation is likely. Thus, the most likely genetic and molecular mechanism of this female Lesch-Nyhan patient is the total HPRT gene deletion on the maternal X chromosome and the rare specific inactivation of the paternal X chromosome.

Hooft et al. (13) described a girl with Lesch-Nyhan syndrome, but her HPRT activity was not determined. Our patient, therefore, must be the first as to a female case with Lesch-Nyhan syndrome evidenced clinically, biochemically and molecularly.

Random inactivation of X chromosome makes the female mosaics of cells with the maternal X active or with the paternal X active. However, individuals who carry a structural anomaly of the X chromosome are exception to this rule. Indeed, if females carry a X-autosome translocation in balanced form, mosaic phenotypes usually do not appear owing to inactivation of the morphologically normal X chromosome in almost all instance (14). Conversely, when the structural anomaly is unbalanced the abnormal X chromosome is generally inactivated. The female Lesch-Nyhan patient is cytogenetically normal and there is no morphological change

TABLE II Summary of Properties of Hybrid Clones

Clones	Chromosomal	Human G6PDH
1	M + P	+
2	M + P	+
3	P	-
4	P	-
5	P	-
6	P	-
1-a	M	+
1-b	P	-
1-c	M + P	+
2-b	P	-
2-c	M + P	+

in X chromosome. Therefore, this patient is a rare case of specific inactivation of morphologically normal X chromosome in the presence of a minor change on the other X chromosome.

IV. SUMMARY AND CONCLUSIONS

HPRT activity and mRNA were undetectable in the cells from a female Lesch-Nyhan patient. Southern blots analysis showed that the patient has only one copy of HPRT gene. Analysis of patient-mouse hybrid cells having human X chromosome(s) indicated a total HPRT gene deletion on the patient's maternal X chromosome and an undetectable change by Southern analysis on the paternal X chromosome. All six independent hybrid clones having the paternal X chromosome alone did not express human glucose-6-phosphate dehydrogenase. Thus, the total HPRT gene deletion on the maternal X chromosome and the specific inactivation of the paternal X chromosome are most likely mechanism of this female Lesch-Nyhan patient.

REFERENCES

1. Lesch, M and Nyhan, W.L., Am.J.Med. 36:561-570 (1964).
2. Seegmiller, J.E., Rosenbloom, F.M. and Kelley, W.N., Science 155:1682-1684 (1967).
3. Kelley, W.N., Rosenbloom, F.M., Henderson, J.F. and Seegmiller, J.E., Proc.Natl.Acad.Sci.USA 57:1753-1739 (1967).
4. Hara, K., Kashiwamata, S., Ogasawara, N., Ohishi, H., Natsume, R., Yamanaka, T., Hakamada, S., Miyazaki, S. and Watanabe, K., Tohok J.Exp.Med. 137:275-282 (1982).
5. Ogasawara, N., Kashiwamata, S., Ohishi, H., Hara, K., Watanabe, K., Miyazaki, S., Kumagai, T. and Hakamada, S., Purine Metabolism im Man-IV. Part A. p.13-18. Plenum, N.Y.(1984).
6. Watson, B., Gormley, I.P., Gradiner, S.E., Evans, H.J. and Harris, H., Exp.Cell Res. 75:401-409 (1972).
7. Bakay, B., Croce, C.M., Koprowski, H. and Nyhan, W.L., Proc.Natl.Acad.Sci.USA 70:1998-2002 (1973).
8. Croce, C.M., Bakay, B., Nyhan, W.L. and Koprowski, H., Proc.Natl.Acad.Sci.USA 70:2590-2594 (1973).
9. Sekiguchi, T. and Sekiguchi, F., Exp.Cell Res. 77: 391-403 (1973).

10. Yang, T.P., Patel, P.I., Chinault, A.C., Stout, J.T., Jackson, L.G., Hildebrand, B.M. and Caskey, C.T., Nature 310:412-414 (1984).
11. Nussbaum, R.L., Crowder, W.E., Nyhan, W.L. and Caskey, C.T., Proc.Natl.Acad.Sci.USA 80:4035-4039 (1983).
12. Boggs, B.B. and Nussbaum, R.L., Somatic Cell Molec.Genet. 10:607-613 (1984).
13. Hooft, C., Van Nevel C. and De Schaepdryer, A.F., Arch.Dis.Child 43:737-747 (1968).
14. Mattei, M.G., Mattei, J.F., Ayme, S. and Giraud, F., Hum.Genet. 61:295-309 (1982).

MOLECULAR APPROACH TO FAMILIAL AMYLOIDOTIC POLYNEUROPATHY: DNA DIAGNOSIS, MOLECULAR PATHOLOGY AND TRANSGENIC MICE

Yoshiyuki Sakaki, Hiroyuki Sasaki, Katsuji Yoshioka, Hirokazu Furuya

Research Laboratory for Genetic Information, Kyushu University, Fukuoka, Japan

Masamitsu Nakazato, Hisayuki Matsuo

Department of Medicine and Biochemistry, Miyazaki Medical School, Miyazaki, Japan

Shigenobu Tone, Yoshihiro Kato

The Mitsubishi-Kasei Institute of Life Sciences, Machida, Japan

I. INTRODUCTION

Familial amyloidotic polyneuropathy (FAP)is a disease showing autosomal dominant inheritance and is clinically characterized by systemic accumulation of amyloid fibrils and progressive disorder of peripheral nerves. The onset of the disease is usually at the age of 30～40 and the disease is highly penetrant. The disease is fatal and it has been desired for a long time to establish a reliable early diagnostic method as well as to develop an effective therapy for the disease.

Biochemical studies revealed a close relationship between FAP and serum transthyretin (TTR, formerly prealbumin) (Costa, et al. 1978). A variant TTR with a single amino acid substitution of methionine for valine at position 30 was identified as a major component of amyloid fibrils from FAP patients of various origins including

Japan, Portugal and Sweden (Tawara et al., 1983; Saraiva et al., 1983; Dwulet and Benson, 1984; Kametani et al., 1984). Based on these findings, we have developed two simple and reliable methods for direct detection of the variant and the mutation responsible for the variant (Nakazato et al., 1984a, Sasaki et al., 1984).

In this paper, we describe molecular analysis of FAP families in Japan to establish early diagnostic methods and to find out a clue to develop an effective therapy

II. DIAGNOSTIC METHODS

A. Diagnosis by Recombinant DNA technique (DNA Diagnosis)

We have cloned a cDNA for normal human TTR and determined its nucleotide sequence (Sasaki et al., 1984). Analysis of the cDNA sequence revealed that the genetic change which could cause the Val→Met substitution at position 30 of TTR leads to formation of new restriction sites for BalI and NsiI:

```
        30                          30
Val Ala Val His    →    Val Ala Met His
GTG GCC GTG CAT         GTG GCC ATG CAT
                           BalI   NsiI
```

The new restriction sites were actually detected in patients of FAP from two major foci of Japan (Arao city and Ogawa village) (Sasaki et al., 1984; Sasaki et al., 1985a). A typical result of the Southern blotting analysis is shown in Fig. 1. It should be noted that the patients are heterozygous for the mutation, which is compatible with the fact that FAP is a dominant inherent genetic disorder.

Nucleotide sequence of TTR cDNA also revealed that a new restriction site for BclI may be formed by the mutation responsible for Jewish type of FAP in which a variant TTR with the Phe→Ile substitution at position 33 was found by Nakazato et al. (1984b):

```
    33                  33
Val Phe Arg    →    Val Ile Arg
GTG TTC AGA         GTG ATC AGA
                       BclI
```

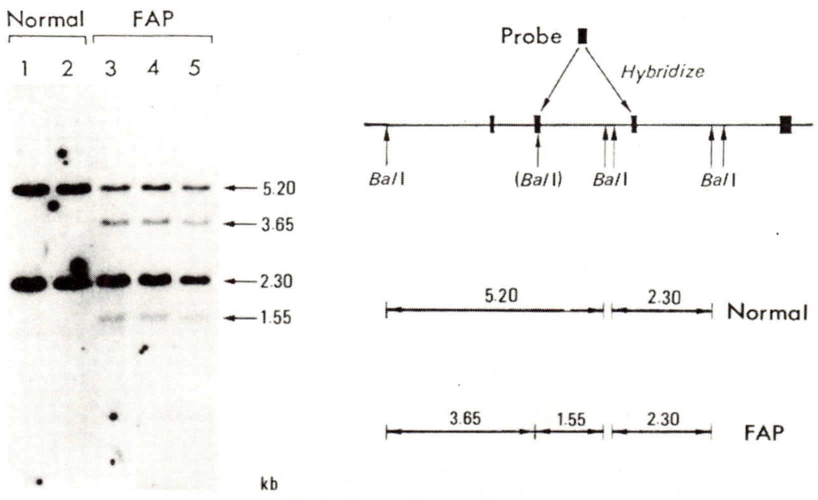

Fig. 1. DNA diagnosis of FAP. DNAs from two normal and three FAP patients were digested by BalI. Southern blotting analysis was carried out as described by Sasaki et al. (1984). Interpretation of the blotting profiles is shematically shown in the right.

No FAP family with the Jewish type of mutation has been found in Japan.

B. Radioimmunoassay (RIA Diagnosis)

Since the variant TTR has a methionine residue at position 30, the variant becomes susceptible to cyanogen bromide at the position 30. Cleavage of the variant TTR by cyanogen bromide coupled with tryptic digestion produces a nonapeptide (positions 22-30) specific for the variant TTR:

Gly-Ser-Pro-Ala-Ile-Asn-Val-Ala-HSer

Antibody against the nonapeptide allowed us to detect the variant TTR in serum by radioimmunoassay technique (Nakazato et al., 1984a). Using the RIA method, the concentration of the variant TTR in patient's serum was estimated to be about half of that of the normal TTR in normal subjects, suggesting that both the normal and mutated gene are expressed co-dominantly in patients.

III ANALYSIS OF FAP FAMILIES

A. Screening of FAP Families in Japan

Using the DNA diagnostic method, number of patients of FAP from various origins has been examined (Yoshioka et al., 1986a). Results are summarized in Table 1. All the patients tested, except one small family case, were found to be carrying the mutation responsible for the Val→Met substitution. Results of diagnosis by the RIA method showed good agreements with those by the DNA method. As far as tested, no homozygote for the mutation was found and no separation between the mutation and FAP.

Table I. Summary of Diagnosis by DNA probe

	positive / subjects
normal control	0 / 15
asymptomatic members of FAP family (beyond the age of onset)	0 / 6
FAP (Arao city)	7 / 7
FAP (Ogawa village)	21 / 21
FAP (Sporadic cases)	4 / 5

Thus, the mutation responsible for the Val→Met substitution is very tightly linked to FAP, and the methods described above should be applicable for presymptomatic and prenatal diagnosis of FAP of various origins and be useful for appropriate genetic advice.

B. Complexity of Disease Process of FAP

Since the onset of FAP is usually at the age of 30-40, a question arose whether the variant TTR is produced long before the age of onset of disease. RIA analysis showed that all the member at the age of less than 20 and carrying the mutation had the variant TTR in serum at the level similar to that of FAP patient (Nakazato et al., 1985)

During the course of screening of FAP families, several very-late onset cases were found. Examples are shown in Fig. 2.

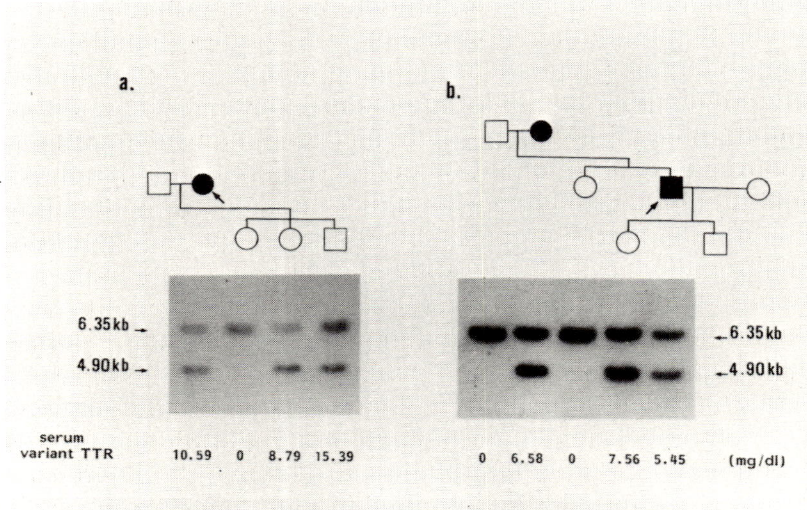

Fig. 2. Analysis of the very late-onset families of FAP. Details are described in the text. Arrows show the probands.

In the first case (Fig. 2a), the patient was a 57-year-old woman, who had the mutation for the Val→Met substitution and the variant TTR in serum and expressed symptoms of typical FAP, but her three children, who were already 34, 32 and 30 years old respectively, did not express any symptoms and signs of FAP, although two of them have the mutation and the variant TTR in serum. In the second case (Fig. 2 b), the patient is a 35-year old man and expressed typical symptoms of FAP. Curiouly, his mother at the age of 56 had the mutation and the variant TTR in serum as her son, but expressed only faint signs of FAP and showed almost no difficulty in ordinary life.

These results suggested that, in addition to the variant TTR, some other genetical and/or evironmental factors may influence the onset and progression of the disease.

IV. STRUCTURAL ANALYSIS OF HUMAN TTR GENE

A. Structure of Normal and Patient TTR Gene

To understand the disease process of FAP more in detail at molecular level, the structure of TTR gene was analyzed. A clone carrying normal human TTR gene was isolated from a human genomic DNA library and its complete nucleotide sequence was determined (Sasaki et al., 1985b). The structure of TTR gene is schematically presented in Fig. 3. The gene has a size of 6.9kb and is composed of four exons and three introns. Some possible regulatory signals were identified: two overlapping sequences homologous to the glucocorticoid-responsive elements (GRE) in the 5'-flanking region, several enhancer-core sequences in introns and the 3' flanking region, and two Alu family sequences with opposite polarity in introns (Fig. 3). Copy number of the TTR gene was estimated to be one per haploid genome by Southern blotting analysis.

Next, the patient type of TTR gene was isolated from a FAP patient and its whole nucleotide sequence determined (Yoshioka et al., 1986a). In comparison with a normal TTR gene sequence, seven base substitutions were found in FAP type TTR gene (See Fig. 3.). No base substitution was found in possible regulatory signals. Among these substitutions, six (substitution I, III-VII) were non-specific, polymorphic changes because these substitutions were found in normal population (Yoshioka et al., 1986b). The substitution II in exon 2 which is responsible for the Val→Met change was found to be only the base substitution

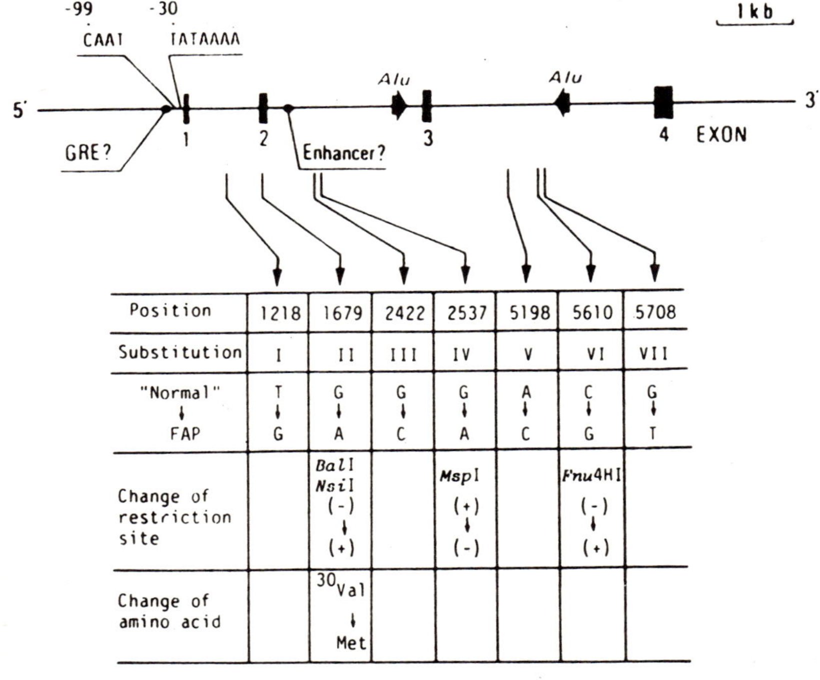

Position	1218	1679	2422	2537	5198	5610	5708
Substitution	I	II	III	IV	V	VI	VII
"Normal" ↓ FAP	T ↓ G	G ↓ A	G ↓ C	G ↓ A	A ↓ C	C ↓ G	G ↓ T
Change of restriction site		*Bal*I *Nsi*I (-) ↓ (+)		*Msp*I (+) ↓ (-)		*Fnu*4HI (-) ↓ (+)	
Change of amino acid		30Val ↓ Met					

Fig. 3. Structure of human TTR gene and base substitutions in a patient TTR gene. Filled boxes are exons. GRE is glucocorticoid responsive element. Figure is drawn based on the results of Sasaki et al. (1985b) and Yoshioka et al. (1985a).

specific for FAP patients. Thus, it becomes clear that FAP is a disease which is primarily caused by qualitative change of serum TTR.

B. Chromosomal Location and RFLPs

The TTR gene was assigned to chromosome 18 using mouse-human hybrid cells (Wallace et al., 1985). To know the linkage of the TTR gene to other chromosomal markers, we tried to determine the precise location of the gene. In collaboration with Sparkes and Simon, the gene was assigned to chromosome 18 q 12.1 using the in situ hybridization (R. Sparkes and M. Simon et al., manuscript in preparation). No gene has been mapped in this region of the chromosome 18, so the TTR gene may become a good marker of this region. RFLPs provide useful information for linkage study. Two RFLPs by MspI and Fnu4HI were found in introns of the TTR gene (Fig.3.) (Yoshioka et al., 1986b). Alelle frequencies (Normal/FAP type) of MspI and Fnu4HI RFLPs were 0.61/0.39 and 0.50/0.50, respectively, in Japanese. These RFLPs may be also useful for haplotype analysis of FAP families. Such studies are now in progress.

V. TRANSGENIC MICE

As described above, FAP is a molecular disorder of TTR but the disease process appears not to be simple. To understand the process of the disease in detail, animal model for the disease should be useful, but no such animal has been reported. Toward the generation of FAP model system, we tried to obtain transgenic mice carrying the human variant TTR gene. Since several genes have been efficiently expressed under the control of mouse metallothionein-I (MT-I) promoter, the variant TTR gene was fused to the MT-I promoter and introduced into fertilized egg of mouse. At present, ten transgenic mice carrying the human variant TTR gene were obtained. Among them three were found to produce significant amount of the variant TTR (Table II). These mice may be useful for studying the biochemical and biophysical nature of the variant TTR and eventually the process of amyloid fibril formation in FAP. Details on these mice will be published elsewhere.

Table II. Transgenic Mice Producing TTR Variant

Mouse	Sex	Human TTR Gene Copies/Diploid	TTR Variant in serum (mg/dl)
43-4	F	0.6	1.40
55-1	M	0.8	1.20
55-2	M	1.2	0.67

ACKNOWLEDGMENTS

We thank Prof. Y. Takagi at Fujita-gakuen Medical School for encouragement, and Drs. R. Sparkes at UCLA Medical Center and M. Simon at California Institute of Technology for communicating us their results prior to publication. We are also grateful to Drs. I. Goto and Y. Kuroiwa at Kyushu University, K. Sahashi at Aichi Medical School, T. Shinoda at Tokyo Metropolitan University, T. Isobe at Kobe University, T. Harada and S. Kito at Hiroshima University, and K. Ichikawa at Hyogo Prefectural Hospital for their cooperation in obtaining blood samples of FAP patients and their families, Ms. N. Yoshioka for technical assistance, and Ms. H. Hamada and Y. Hayashi for assistance in preparation of the manuscript. This work is supported in part by grants from the Ministry of Education, Science and Culture and from Foundation for Promotion of Cancer Research granted by Japan Ship-building Industry Foundation

REFERENCES

1. Costa, P. P., Figueira, A. S. and Bravo, F. R. (1978). Proc. Natl. Acad. Sci. USA, 75: 4499
2. Dwulet, F. E. and Benson, M. D. (1984). Proc. Natl. Acad. Sci. USA, 81: 694
3. Kametani, F., Tonoike, H., Hoshi, A., Shinoda, T. and Kito, S. (1984). Biochem. Biophys. Res. Commun. 125: 622
4. Nakazato, M., Kangawa, K., Minamino, N.,

Tawara ,S., Matsuo, H. and Araki, S. (1984a). Biochem. Biophys. Res. Commun. 122: 719
5. Nakazato, M., Kangawa, K., Minamino, N., Tawara, S., Matsuo, H. and Araki, S. (1984b) Biochem. Biophys. Res. Commun. 123:921
6. Nakazato, M., Kurihara, T., Kangawa, K. and Matsuo, H. (1985). Lancet 1985i: 99
7. Saraiva, M. J. M., Costa, P.P., Birken, S. and Goodman, D. S. (1983). Trans. Ass. Amer. Phys. 96:261
8. Sasaki, H., Sakaki, Y., Matsuo, H., Goto, I., Kuroiwa, Y. Sahashi, I., Takahashi, A., Shinoda, T., Isobe, T. and Takagi, Y. (1984). Biochem. Biophys. Res. Commun. 125: 636
9. Sasaki H., Sakaki, Y., Takagi, Y., Sahashi, K., Takahashi, A., Isobe, T., Shinoda, T., Matsuo, H., Goto, I. and Kuroiwa, Y. (1985a). Lancet i: 100
10. Sasaki, H., Yoshioka, N., Takagi, Y. and Sakaki, Y. (1985b). Gene 37:191
11. Tawara, S., Nakazato, M., Kangawa, K., Matsuo, H. Araki, S. (1983). Biochem. Biophys. Res. Commun. 116: 880
12. Yoshioka, K., Sasaki, H., Yoshioka, N., Furuya, H., Harada, T., Kito, S. and Sakaki, Y. (1986a). Mol. Biol. Med. (in press)
13. Yoshioka, K., Yoshioka, N., Nakabeppu, K. and Sakaki, Y. (1986b). Nucleic. Acids. Res. 14:3147
14. Wallace, M. R., Naylor, S. L., Kluve-Beckerman, B. Long, G.L., McDonald, L., Shows, T. B., and Benson, M. D. (1985). Biochem. Biophys. Res. Commun. 129: 753

CONVERSION OF NON-RESPONDER MICE TO GLϕ INTO RESPONDERS BY THE TRANSFER OF THE Eα GENE[1]

Hitoshi Kikutani, Ken-ichi Yamamura[2], Masao Kimoto[3] and Tadamitsu Kishimoto

Institute for Molecular and Cellular Biology, Osaka University, Suita City, Osaka Japan

I. INTRODUCTION

Induction of the antibody response to antigen challenge requires the collaboration between T and B lymphocytes. Without T cell activation, for a given antigen, B cell differentiation into antibody secreting cells does not occur. Antigens are processed by monocytes and are presented to T cells in a context of self-Ia complex. T cells can recognize foreign antigens only in a context of self-Ia complex and are activated. Therefore, Ia molecules are essential for the induction of the immune response. Ia molecules are heterodimeric glycoproteins and composed of α and β chains; I-A is composed of Aα and Aβ and I-E of Eα and Eβ.

C57BL/6 mice do not express I-E molecules, since the 5' portion of the $E\alpha^b$ gene including the transcriptional promoter and exon 1 is deleted (1), and these mice are not responsive to a certain set of antigens, such as a synthetic polypeptide, glutamyl-lysyl-phenylalanine (GLϕ) (2,3). In this study, we attempted to introduce the intact $E\alpha^d$ gene,

---

1) Supported by a grant from the Ministry of Education, Science and Culture.

2) Present address: Department of Genetics, School of Medicine, Kumamoto University.

3) Present address: Department of Immunology, Saga Medical College

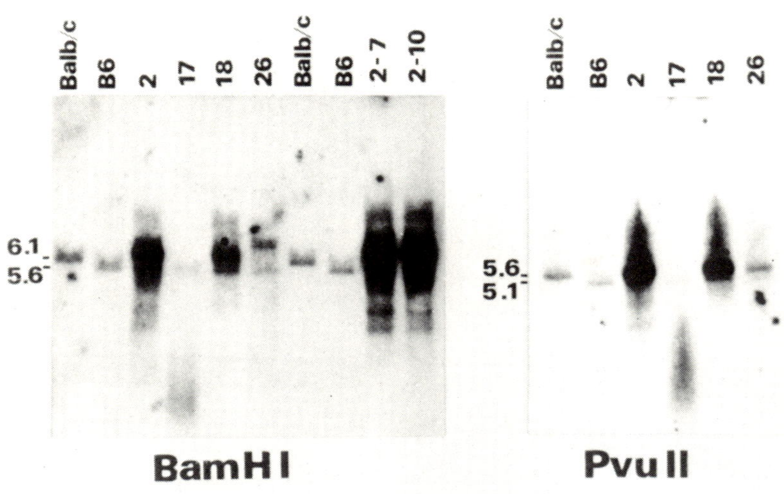

Fig.1. Southern blot analysis of the Eα gene in transgenic mice.

which had been cloned from BALB/c mice, into C57BL/6 mice and explored whether the immune responsiveness to GLϕ can be recovered in the transgenic C57BL/6 mice with the $E\alpha^d$ gene (4).

II. PREPARATION OF TRANSGENIC MICE WITH $E\alpha^d$ GENE

Approximately 200 copies of the 14kb SacII-XhoI fragment containing the entire $E\alpha^d$ gene sequence were microinjected into each fertilized egg of the C57BL/6 mouse. In total, 30 mice were born. These mice were partially hepatectomized and the DNA isolated from the livers were analysed by Southern blotting to determine whether the injected $E\alpha^d$ gene is retained in the host DNA. Because of the deletion of the 5' portion of the $E\alpha^d$ gene, BALB/c DNA yielded a DNA fragment 0.5kb larger than that of C57BL/6 DNA when these DNAs were digested either with BamHI or PvuII (1). This allowed us to distinguish the injected $E\alpha^d$ gene from the endogeneous defective $E\alpha^b$ gene present in the host mice. As shown in Fig.1, DNA from transgenic mice 2 and 18 gave the 6.1kb BamHI band and 5.6kb PvuII band diagnostic of

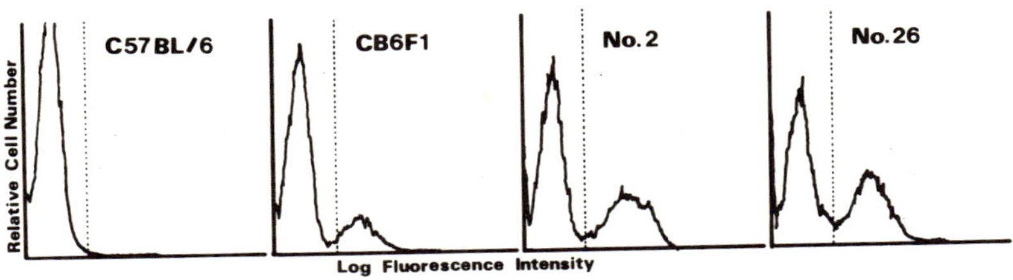

Fig.2. Expression of the I-$E^{d/b}$ antigens on lymphocytes.

the $E\alpha^d$ gene, indicating that these mice have retained the injected $E\alpha^d$ gene. DNA from the transgenic mouse 26 displayed the 5.6kb PvuII band but the size of the corresponding BamHI fragment was somewhat greater than 6.1kb. This discrepancy may result from loss of the 5' BamHI site. From an approximate calculation based on the intensities of the $E\alpha^d$ bands, mice 2 and 18 seem to have at least 10 copies of the $E\alpha^d$ gene. These $E\alpha^d$ genes were transmitted to a fraction of offsprings, two examples of which (nos. 2-7 and 2-10) are shown in Fig.1.

III. TISSUE SPECIFIC EXPRESSION OF THE INJECTED $E\alpha^d$ GENE

The expression of the $E\alpha^d$ gene in transgenic mice was studied by the surface staining of peripheral blood mononuclear cells with a monoclonal anti-I-E, Y-17 (5). As shown in Fig.2, about one-third of the peripheral lymphocytes from transgenic mice were stained with Y-17, indicating that the injected $E\alpha^d$ genes were expressed, their products could pair with the endogeneous $E\beta^b$ chains and $E\alpha^dE\beta^b$ molecules were expressed on the surface of the lymphocytes of transgenic mice. In order to study whether

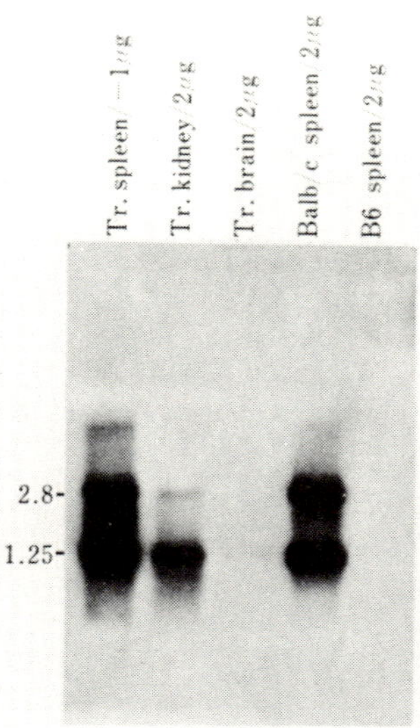

Fig.3. Northern blot analysis of Eα mRNA from various tissues of a transgenic mouse.

I-E molecules are expressed on B cells, the lymphocytes were stained with anti-IgM and Y-17 monoclonal antibody. Two colour FACS analysis demonstrated that I-E molecules were expressed on IgM^+ B cells but not on IgM^- cells which were mainly T cells. The expression of I-E molecules on monocytes was also studied. A small fraction of peritoneal monocytes in $CB6F_1$ mice expressed I-E molecules and the stimulation with γ-interferon induced I-E expression on most of the peritoneal monocytes. In contrast, most of the peritoneal monocytes from the transgenic mice expressed I-E molecules in the absence of γ-interferon stimulation. The reason for this constitutive expression is unknown, but it may be related to the fact that the transgenic mice carry a large copy number of the $E\alpha^d$ genes. In order to examine the transcription of the microinjected $E\alpha^d$ gene, a Northern blot analysis of the Eα transcript in various tissues of the transgenic mice was performed. As shown in Fig.3, a relatively large amount of Eα mRNA of 1.25kb was observed in poly$(A)^+$ RNA from spleen and kidney. Similar amounts of Eα transcripts were also observed in BALB/c kidneys. However, the expression of I-E molecules in kidneys was not observed by the tissue-staining with a peroxidase-conjugated monoclonal antibody. The presence of Eα transcripts in kidneys

TABLE I. Stimulation of $I\text{-}E^{b/d}$ specific T cell clone with spleen cells from the transgenic mice

Responder cells	Stimulater cells	^{3}H-TdR uptake
G.D 2.2 clone	B6	4,741 ± 276
($I\text{-}E^{b/d}$ specific)	$CB6F_1$	43,373 ± 5,885
	transgenic	41,867 ± 2,219

might be due to the contamination of blood.

IV. IMMUNOLOGICAL FUNCTIONS OF I-E MOLECULES EXPRESSED IN THE TRANSGENIC MICE

In order to study whether $E\alpha^{d}E\beta^{b}$ molecules expressed on the lymphocytes in the transgenic mice can be recognized as alloantigens by T cells, mixed lymphocyte reaction was performed. Spleen cells from the transgenice mice could induce a significant proliferation of T cells from C57BL/6 mice. The result was further confirmed by utilizing $E\alpha^{d}E\beta^{b}$-specific autoreactive T cell clone, which had been established from KLH-primed T cells from $CB6F_1$ mice by the method of Kimoto and Fathman. As shown in Table 1, an $E\alpha^{d}E\beta^{b}$-specific T cell clone showed a significant proliferation by stimulation with spleen cells from the transgenic mice but not from C57BL/6 mice.

Finally, we examined whether the immune response against GLϕ could be restored in the transgenic mice. The transgenic mice as well as C57BL/6 mice were immunized with GLϕ included in complete Freund's adjuvant and their spleen cells were stimulated in vitro with GLϕ. As shown in Table 2, GLϕ could induce significant proliferation of the lymphocytes from the transgenic mice but not of the lympohcytes from C57BL/6 mice. The results suggest strongly that the product of the microinjected Eα gene is expressed on the surface of the antigen-presenting cells of the host mouse in such a way as to allow the restoration of responsiveness to the antigen, GLϕ.

TABLE II A transgenic mice aquires responsiveness to GLϕ

GLϕ-immunized cells from	Antigens		
	(-)	GLϕ	PPD
C57BL/6	6,445 ± 1,268	6,236 ± 837	47,296 ± 13,307
Transgenic	4,434 ± 438	15,228 ± 4,612	20,314 ± 5,061

V. CONCLUSION

In the present study, we produced transgenic mice which express functional I-E antigens by microinjecting the cloned $E\alpha^d$ DNA into fertilized eggs of C57BL/6 mice. The transgenic mice tissue-specifically expressed I-E molecules on B cells and monocytes. These I-E molecules could be recognized by antibodies as well as T cells as alloantigens. The defective immune response against GLϕ in C57BL/6 mice was restored in the transgenic mice, directly demonstrating that the Eα gene is a gene controlling the immune response. It is possible to establish new strains carrying the $E\alpha^d$ gene by crossing these mice with other B6 or B10 congenic mice. Thus, these transgenic mice should be a powerful tool for the analysis of the effect of well-defined class II gene on immune response.

ACKNOWLEDGEMENTS

We thank Ms. K. Kubota, J. Mori, and M. Kawata for their excellent secretarial assistance.

REFERENCES

1. Mathis, D.J., Benoist, C., Williams, V.E.II., Kanter, M.R. and McDevitt, H.O. Proc. Natl. Acad. Sci. U.S.A. 80:273 (1983).
2. Benacerraf, B. and Dorf, M.E. Cold Spring Harb. Symp. quant. Biol. 41:465 (1976).
3. Schwartz, R.H., David, C.S., Sachs, D.H. and Paul, W.E. J. Immunol. 117:531 (1976).
4. Yamamura, K., Kikutani, H., Folsom, V., Clayton, L.K.,

Kimoto, M., Akira, S., Kashiwamura, S., Tonegawa, S. and Kishimoto, T. Nature 316:67 (1985).

5. Lerner, E.A., et al. J. Exp. Med. 152:1085 (1980).

MOLECULAR BASIS OF UREA CYCLE DISORDERS

Masataka Mori
Masaki Takiguchi
Yoshihiro Amaya
Yougo Haraguchi

Institute for Medical Genetics
Kumamoto University Medical School
Kumamoto

Susumu Kawamoto

Department of Bacteriology
Teikyo University School of Medicine
Tokyo

Akira Ohtake

Department of Pediatrics
Chiba University School of Medicine
Chiba

I. INTRODUCTION

The urea cycle is the major pathway for detoxication of ammonia which is formed in amino acid metabolism. The urea cycle classically involves five enzymes, carbamyl phosphate synthetase (CPS), ornithine transcarbamylase (OTC), argininosuccinate synthetase (AS), argininosuccinate lyase (AL) and arginase (ARG). N-Acetylglutamate synthetase (AGS) which is responsible for synthesis of N-acetylglutamate, an allosteric and essential activator of CPS, may be involved as a member of the urea cycle enzymes. Among these six enzymes of the urea synthetic pathway, the first three enzymes (AGS, CPS and OTC) are localized in the mitochondrial matrix, whereas the last three enzymes (AS, AL and ARG) are localized in the cytosol. Enzyme deficiencies have been known for all the enzymes of the urea cycle. If one of these enzymes is missing, conversion of ammonia to urea is impaired, and hyperammonemia occurs. In this paper, we report molecular analysis of several cases of inherited urea cycle disorders and isolation of cDNA clones for OTC, AL and ARG.

II. RESULTS AND DISCUSSION

A. CPS Deficiency

CPS deficiency is relatively rare in inherited urea cycle disorders. Although data on several patients with this disorder have been reported, there is no report which refers to the molecular mechanism responsible for CPS deficiency. We analyzed two siblings with the deficiency who died at the age of about 40 days (1).

In neither patient was there detectable activity of CPS in the liver. CPS protein can be readily detected by electrophoresis of human liver homogenates after denaturation with SDS. CPS protein was missing in the liver of either patient by SDS gel electrophoresis of the liver homogenates. The immunoreactive protein was determined by single radial immunodiffusion and immunoblotting, and no cross-reactive material was detected in these patients.

For elucidation of the molecular mechanism responsible for the defect of CPS protein in the liver of these patients, we determined the level of translatable mRNA for CPS by in vitro protein synthesis using the method described previously (2). Total RNA extracted from the livers of controls and one of the patients were translated in a rabbit reticulocyte lysate system and in vitro-synthesized CPS precursor (3,4) was isolated by immunoprecipitation and SDS gel electrophoresis, and was visualized by fluorography. The CPS precursor was detected with the control liver mRNAs, but not with the patient liver mRNA. These results indicate that the translatable mRNA is lacking in the patient liver, and that the absence of enzyme protein in the patient is at least mainly due to decreased synthesis. Two very similar cases of CPS deficiency have been observed (5,6). These deficiencies could be due to a mutation affecting either transcription of the gene, processing of the mRNA precursor, stability of mRNA or its translatability. Studies with cDNA probes (7,8) should reveal the molecular mechanism responsible for these disorders.

B. OTC Deficiency

OTC deficiency, an X-linked trait, is one of the more common disorders of the urea cycle and is characterized by hyperammonemia with orotic aciduria. We have examined enzyme activity, amount of the enzyme protein, and in some cases, mRNA activity of eight Japanese patients with OTC deficiency. The results are summarized in TABLE I.

Enzyme activity in the liver samples of the patients were less than nine % of control except for the patient 8 (32% of control). Among the eight patients, only the patient 4 had a normal amount of immunoreactive enzyme protein (10). Kinetic properties of the enzyme from the patient appeared to be identical to those of control enzyme. Isoelectric focusing of the patient enzyme gave a major peak of immunoreactive protein which was catalytically inactive, and a minor peak of the protein which was catalytically active. These results indicate that the patient's liver contained an inactive form of OTC in addition to an active form of the enzyme. The active form of the enzyme might be derived from the normal OTC gene on the X chromosome, whereas the inactive form might be derived from the mutant gene.

Three patients with highly decreased amount of the enzyme protein (patients 3, 5 and 6) were assayed for OTC mRNA activity. The mRNA activity was markedly decreased in patients 5 and 6, and we concluded that the decreased amount of the enzyme in these two patients is mainly due to decrease synthesis (11). The mRNA from one of the patients

TABLE I. OTC Deficiency

Case	Sex	Enzyme activity (%)	Amount of enzyme protein	mRNA activity	Reference
1	Male	0	0		Unpublished
2	Male	0	0		Unpublished
3	Male	0	0	Normal	(9)
4	Female	9	Normal		(10)
5	Female	4	Highly decreased	Highly decreased	(11)
6	Female	4	Highly decreased	Highly decreased	(11)
7	Female	5	Highly decreased		Unpublished
8	Female	32	Decreased		Unpublished

directed the synthesis of a truncated OTC precursor (30,000 daltons) in addition to a normal precursor (40,000 daltons). Again, the normal precursor might be derived from the normal gene, whereas the truncated precursor might be derived from abnormal gene.

We have especially interested in patient 3. The patient, who died at the age of 3 days, had no OTC activity and no detectable immunoreactive protein. However, OTC mRNA activity in the patient's liver, measured by in vitro translation, was similar to or even higher than those in controls. Therefore, it is likely that the OTC precursor was synthesized normally in the patient's liver. OTC is initially synthesized as a larger precursor in the cytosol of liver cells and is transported into the mitochondria and processed to the mature form of the enzyme (12-15). The possibility that the absence of OTC protein in the patient's liver is due to impaired transport or processing of the precursor, was tested in a cell-free reconstitution system in which the OTC precursor synthesized in vitro can be taken up and processed to the mature enzyme. Because human liver mitochondria could not be easily prepared, and because the transport and processing system is not species-specific (15), we used rat liver mitochondria. The OTC precursor synthesized with mRNA from the patient's liver as well as from control livers could be taken up by rat liver mitochondria and processed to the mature enzyme. These results suggest that the step of the precursor transport and processing was not impaired in the patient's liver. It is likely that the maturated OTC or its precursor was unstable and degraded rapidly in the patient's liver, although there are many other possibilities.

C. OTC Deficiency in Mice

Two mouse OTC mutations, spf (sparse-fur) and spf-ash (sparse-fur with abnormal skin and hair) have been known (16,17). Hepatic OTC activity in spf mice is reduced to a few % at pH 7.4, the pH optimum is shifted from 8.5 to 9.5, and the amount of the enzyme protein is somewhat increased. On the other hand, activity and amount of the enzyme in spf-ash mice are 5-10% of control, and kinetic properties are normal (18). Thus, the deficiency in spf mice is qualitative type, whereas that in spf-ash mice is quantitative type.

We showed that OTC mRNA activity in spf mice, measured by in vitro translation, is reduced to 60% of control (19). The increased amount of enzyme protein is presumably attributable to a diminution of the enzyme degradation. The mRNA activity in spf-ash mice is reduced to about 10% of control, indicating that the decreased enzyme amount is due to decreased synthesis (19).

The subunit molecular weight of the spf enzyme precursor synthesized in vitro is practically the same as that of the normal enzyme precursor (19). On the other hand, in the case of spf-ash enzyme, we observed two discrete in vitro products on SDS-gel electrophoresis; one comigrated with the normal precursor and the other moved slightly slower (19). Both products appeared to be taken up and processed to the mature form of the enzyme. Similar results were reported by Rosenberg and others (20).

We performed RNA blot hybridization using a cloned rat cDNA as a probe. No difference in mRNA size (about 1.8 kilobases) was detected between control, spf and spf-ash mice. The amounts of OTC mRNA in spf and spf-ash mice were 70% and 10% of control, respectively. These values agree well with the translatable mRNA levels determined by in vitro translation. These results are summarized in TABLE II. The spf-ash mRNA(s) directed the formation of two distinct OTC precursors, but only a single mRNA band was detected by blot analysis.

To investigate the size and the amount of OTC mRNA precursors in the mutant mice, we performed blot hybridization analysis of nuclear RNA (21). Nuclear RNA from spf mouse liver contained almost the same amount of hybridizable OTC mRNA as that of normal mouse liver. In contrast, amount of nuclear RNA for OTC in spf-ash mouse liver was much reduced compared with that in normal mouse liver, and no accumulation of mRNA precursors was found.

Taking together, the mutation in spf mice is almost certainly a point mutation which affects the enzyme activity. On the other hand, the deficiency in spf-ash mice could be due to a mutation affecting either transcription of the OTC gene or mRNA stability.

TABLE II. OTC Deficiency Mice

Mouse	Enzyme activity (% of normal)		Enzyme protein (% of normal)	mRNA Activity (% of normal)	Hybridizable mRNA	
	pH 7.4	pH 9.5			Cytoplasm (% of normal)	Nucleus
Spf	10	150	150	58	60	Normal
Spf-ash	7	7	10	10	10	Decreased

D. Cloning and Nucleotide Sequence of Rat OTC cDNA

Messenger RNA of rat OTC was enriched by immunoprecipitation of rat liver free polysomes and cDNA clones were isolated. One of the clones containing a 1.6 kilobase insert was subjected to nucleotide sequence analysis. The deduced amino acid sequence indicates that the OTC precursor consists of the mature enzyme of 322 amino acid residues and an amino-terminal peptide extension (presequence) of 32 amino acid residues (22)(FIGURE 1). The presequence contains 8 basic amino acid residues, no acidic residue, and no hydrophobic amino acid stretch. The amino acid sequence of the rat OTC precursor was compared with the sequence of human enzyme precursor reported by Horwich et al. (23). The sequences of the mature enzyme portions are 93% identical, whereas those of the presequences are 67%. Several lines of evidence indicates that the presequence contains a sufficient information for mitochondrial targetting of the precursor and its transport into the organelle.

There is a highly conserved segments in the mature portions of the rat and human enzymes, and *Escherichia coli* enzyme (24), corresponding in the rat and human enzymes to the residues 53 to 68 (FIGURE 2). These segments are also highly homologous to a segment of *E. coli* aspartate transcarbamylase catalytic subunit (25,26). Since OTC and aspartate transcarbamylase use carbamyl phosphate as a common substrate, these segments presumably represent binding sites for carbamyl phosphate.

E. Cloning of cDNAs for Rat Argininosuccinate Lyase and Rat and Human Arginase

Argininosuccinate lyase (AL) and arginase (ARG) catalyze the last two steps of urea synthesis in the liver of ureotelic animals. AL from bovine and human liver consists of four identical subunits of about 50,000 daltons. ARG from rat liver is trimer or tetramer of identical subunits of about 40,000 daltons. Deficiency of AL leads to argininosuccinic aciduria and that of ARG to argininemia.

A cDNA bank of rat liver constructed with the plasmid expression vector pUC8, was screened immunologically by using antibodies against AL and ARG. Four positive clones of AL cDNA (27) and three positive clones of ARG cDNA (28) were isolated from about 60,000 transformants. An AL cDNA clone contained an insert of about 1.5 kilobase pairs and an ARG cDNA clone contained an insert of about 1.35 kilobase pairs. The cDNA clones were identified by hybrid-selected translation. A transformant having an AL cDNA plasmid

```
-58                                                            GTGCCTGCCTGCGGAACTCTCTAGACCATAGATTCCTCCTCCACTCTAGCAAGAGAAG    -1

1  ATG CTG TCT AAT TTG AGG ATC CTG CTC AAC AAG GCA GCT CTT AGA AAG GCT CAC ACT TCC ATG GTT CGA AAT TTT CGG TAT GGG AAG CCA    90
1  Met Leu Ser Asn Leu Arg Ile Leu Leu Asn Lys Ala Ala Leu Arg Lys Ala His Thr Ser Met Val Arg Asn Phe Arg Tyr Gly Lys Pro    30
           Phe                             Asn         Phe     Asn Gly     Asn Phe                         Cys     Gln

   GTC CAG AGT CAA GTA CAG CTG AAA GGC CGT GAC CTC CTC ACC CTG AAG AAC TTC ACA GGA GAG GAG ATT CAG TAC ATG CTA TGG CTC TCT   180
   Val Gln Ser Gln Val Gln Leu Lys Gly Arg Asp Leu Leu Thr Leu Lys Asn Phe Thr Gly Glu Glu Ile Gln Tyr Met Leu Trp Leu Ser    60
   Leu    ↑Asn Lys

   GCA GAT CTG AAA TTC AGG ATC AAA CAG AAA GGA GAA TAC TTG CCT TTA TTG CAA GGG AAA TCC TTA GGG ATG ATT TTT GAG AAA AGA AGT   270
   Ala Asp Leu Lys Phe Arg Ile Lys Gln Lys Gly Glu Tyr Leu Pro Leu Leu Gln Gly Lys Ser Leu Gly Met Ile Phe Glu Lys Arg Ser    90

   ACT CGA ACA AGA CTG TCC ACA GAA ACA GGC TTC GCT CTT CTG GGA GGA CAT CCT TCT TTT CTT ACC ACA CAA GAC ATT CAC TTG GGC GTG   360
   Thr Arg Thr Arg Leu Ser Thr Glu Thr Gly Phe Ala Leu Leu Gly Gly His Pro Ser Phe Leu Thr Thr Gln Asp Ile His Leu Gly Val   120
                                                                           Cys     Pro

   AAT GAA AGT CTC ACA GAC ACA GCT CGT GTG TTA TCT AGC ATG ACA GAT GCA GTG TTA GCT CGA GTG TAT AAA CAA TCA GAT CTG GAC ATC   450
   Asn Glu Ser Leu Thr Asp Thr Ala Arg Val Leu Ser Ser Met Thr Asp Ala Val Leu Ala Arg Val Tyr Lys Gln Ser Asp Leu Asp Ile   150
                                                           Ala                                                         Thr

   CTG GCT AAG GAA GCA ACC ATC CCA ATT GTC AAC GGA CTG TCA GAC CTG TAT CAT CCT ATC CAG ATC CTG GCT GAT TAC CTT ACA CTC CAG   540
   Leu Ala Lys Glu Ala Thr Ile Pro Ile Val Asn Gly Leu Ser Asp Leu Tyr His Pro Ile Gln Ile Leu Ala Asp Tyr Leu Thr Leu Gln   180
                       Ser             Ile

   GAA CAC TAT GGC TCT CTC AAA GGT CTC ACC CTC AGC TGG ATA GGA GAT GGG AAC AAT ATC CTG CAC TCC ATC ATG ATG AGT GCT GCA AAA   630
   Glu His Tyr Gly Ser Leu Lys Gly Leu Thr Leu Ser Trp Ile Gly Asp Gly Asn Asn Ile Leu His Ser Ile Met Met Ser Ala Ala Lys   210
               Ser                                 Cys Phe

   TTC GGG ATG CAC CTT CAA GCA GCT ACT CCA AAG GGT TAT GAG CCA GAT CCT AAT ATA GTC AAG CTA GCA GAG CAG TAT GCC AAG GAG AAT   720
   Phe Gly Met His Leu Gln Ala Ala Thr Pro Lys Gly Tyr Glu Pro Asp Pro Asn Ile Val Lys Leu Ala Glu Gln Tyr Ala Lys Glu Asn   240
                                                                   Ala Ser Val Thr

   GGT ACC AGG TTG TCA ATG ACA AAT GAT CCA CTG GAA GCA GCA CGT GGA GGC AAT GTA TTA ATT ACA GAT ACT TGG ATA AGC ATG GGA CAA   810
   Gly Thr Arg Leu Ser Met Thr Asn Asp Pro Leu Glu Ala Ala Arg Gly Gly Asn Val Leu Ile Thr Asp Thr Trp Ile Ser Met Gly Gln   270
           Lys     Leu Leu                                 His                                                         Arg

   GAG GAT GAG AAG AAA AAG CGT CTT CAA GCT TTC CAA GGT TAC CAG GTT ACA ATG AAG ACT GCT AAA GTG GCT GCG TCT GAC TGG ACG TTT   900
   Glu Asp Glu Lys Lys Lys Arg Leu Gln Ala Phe Gln Gly Tyr Gln Val Thr Met Lys Thr Ala Lys Val Ala Ala Ser Asp Trp Thr Phe   300
       Glu

   TTA CAC TGC TTG CCT AGA AAG CCA GAA GAA GTA GAT GAT GAA GTG TTT TAT TCT CCG CGG TCA TTA GTG TTC CCA GAG GCA GAA AAT AGA   990
   Leu His Cys Leu Pro Arg Lys Pro Glu Glu Val Asp Asp Glu Val Phe Tyr Ser Pro Arg Ser Leu Val Phe Pro Glu Ala Glu Asn Arg   330

   AAG TGG ACA ATC ATG GCT GTC ATG GTA TCC CTG CTG ACA GAC TAC TCA CCT GTG CTC CAG AAG CCA AAG TTC TGATGCCTGTCAAGAGGACGAAA  1085
   Lys Trp Thr Ile Met Ala Val Met Val Ser Leu Leu Thr Asp Tyr Ser Pro Val Leu Gln Lys Pro Lys Phe                          354
                                                                       Gln
   AACCCAAAAGACAAAAAAATCTGTTCTTTAGCAGCAGAATAAGTCAGTTTATGTAGAAAAGAGAAGAATTGAAATTGTAAACACATCCCTAGTGCGTGATATAATTATGTAATTGCTTT  1204

   GCTATTGTGAGAATTGCTTAAAGCT                                                                                                1229
```

FIGURE 1. (Continued to the next page)

FIGURE 1. Nucleotide and corresponding amino acid sequence of rat OTC cDNA (22). Nucleotides are numbered in the 5' to 3' direction, beginning with the first residue of the ATG triplet encoding the initiator methionine, and the nucleotides on the 5' side of residue 1 are indicated by negative numbers. Deduced amino acid sequence is indicated below the nucleotide triplets. Arrowhead indicates the cleavage site and mature polypeptide begins at Ser-33. Predicted amino acids of human OTC precursor (23) different from those of the rat enzyme precursor are shown below the rat sequence.

E. coli OTC	50	I	F	E	K	D	S	T	R	T	R	C	S	F	E	V	65
Rat OTC	53	I	F	E	K	R	S	T	R	T	R	L	S	T	E	T	68
Human OTC	53	I	F	E	K	R	S	T	R	T	R	L	S	T	E	T	68
E. coli ATC	47	C	F	F	E	A	S	T	R	T	R	L	S	F	Q	T	62

FIGURE 2. Comparison of amino acid sequence of a portion of rat (22) and human (23) OTC with the sequence of a portion of E. coli OTC (24) and E. coli aspartate transcarbamylase (ATC) catalytic subunit (25,26).

---

expressed a specific product of 25,000 daltons, but no AL activity was detected. On the other hand, a tranformant having an ARG cDNA plasmid expressed a specific product of 43,000 daltons, and a high ARG activity was detected in this transformant.

Northern blot analysis of RNA from rat liver, small intestine, kidney, spleen and heart was carried out. The AL mRNA of about 2.1 kilobases long was detected in the liver and in a lesser amount in kidney and spleen, but not in the small intestine and heart of the rats. The tissue distribution of the mRNA is very similar to that of the enzyme activity. On the other hand, the ARG mRNA of about 1.6 kilobases long was detected in the liver, but not in the small intestine, kidney, spleen and heart of the rats. Again, the tissue distribution of the mRNA is consistent with that of ARG activity.

A human liver cDNA bank constructed with λgt11 was screened with a rat ARG cDNA as a probe, and several positive clones have been isolated. One of the clones had an insert of about 1.4 kilobase pairs. Nucleotide sequencing of the rat and human ARG cDNAs is under way. These cDNA sequences will facilitate further studies on genetic analysis, DNA diagnosis and gene therapy of the enzyme deficiencies.

REFERENCES

1. Ohtake, A., Miura, S., Mori, M., Takayanagi, M., Kakinuma, H., Tatibana, M. and Nakajima, H., Acta Pediatr. Jpn 26:262 (1984)
2. Mori, M., Miura, S., Tatibana, M. and Cohen, P. P., J. Biol. Chem. 256:4127 (1981)
3. Mori, M., Miura, S., Tatibana, M. and Cohen, P. P., Proc. Natl. Acad. Sci. USA 76:5071 (1979)
4. Shore, G. C., Carignan, P. and Raymond, Y., J. Biol. Chem. 254:3141 (1979)
5. Graf, L., Hoogenraad, N., Brown, G. and Haan, E. A., J. Inher. Metab. Dis. 7:104 (1984)
6. Suzuki, Y., Matsushima, A., Ohtake, A., Mori, M., Tatibana, M. and Orii, T., Eur. J. Pediatr. in press
7. Adcock, M. and O'Brien, W. E., J. Biol. Chem. 259:13471 (1984)
8. Nyunoya, H., Broglie, K. E., Widgren, E. E. and Lusty, C. J., J. Biol. Chem. 260:9346 (1985)
9. Saheki, T., Imamura, Y., Inoue, I., Miura, S., Mori, M., Ohtake, A., Tatibana, M., Katsumata, N. and Ohno, T., J. Inher. Metab. Dis. 7:2 (1984)
10. Mori, M., Uchiyama, C., Miura, S., Tatibana, M. and Nagayama, E., Clin. Chim. Acta 104:291 (1980)
11. Kodama, H., Ohtake, A., Mori, M., Okabe, I., Tatibana, M. and Kamoshita, S., J. Inher. Metab. Dis. in press
12. Mori, M., Miura, S., Tatibana, M. and Cohen, P. P., J. Biochem. 88:1829 (1980)
13. Mori, M., Miura, S., Morita, T. and Tatibana, M., Mol. Cell. Biochem. 49:97 (1982)(Review)
14. Conboy, J. G., Kalousek, F. and Rosenberg, L. E., Proc. Natl. Acad. Sci. USA 76:5724 (1979)
15. Morita, T., Miura, S., Mori, M. and Tatibana, M., Eur. J. Biochem. 122:501 (1982)
16. DeMars, R., Levan, S. L., Trend, B. L. and Russell, C. B., Proc. Natl. Acad. Sci. USA 73:1693 (1976)
17. Quereshi, I. A., Letarte, J. and Quellet, R., Pediat. Res. 13:807 (1979)
18. Briand, P., Francois, B., Rabier, D. and Cathelineau, L., Biochim. Biophys. Acta 704:100 (1982)
19. Briand, P., Miura, S., Mori, M., Cathelineau, L., Kamoun, P. and Tatibana, M., Biochim. Biophys. Acta 760:389 (1983)
20. Rosenberg, L. E., Kalousek, F. and Orsulak, M. D., Science 222:426 (1983)
21. Ohtake, A., Takayanagi, M., Yamamoto, S., Kakinuma, H., Nakajima, H., Tatibana, M. and Mori, M., J. Inher. Metab. Dis. in press

22. Takiguchi, M., Miura, S., Mori, M., Tatibana, M., Nagata, S. and Kaziro, Y., Proc. Natl. Acad. Sci. USA 81:7412 (1984)
23. Horwich, A. L., Fenton, W. A., Williams, K. R., Kalousek, F., Kraus, J. P., Doolittle, R. F., Konigsberg, W. and Rosenberg, L. E., Science 224:1068 (1984)
24. Bencini, D. A., Houghton, J. E., Hoover, T. A., Foltermann, K. F., Wild, J. R. and O'Donovan, G. A., Nucleic Acids Res. 11:8509 (1983)
25. Hoover, T. A., Roof, W. D., Foltermann, K. F., O'Donovan, G. A., Bencini, D. A. and Wild, J. R., Proc. Natl. Acad. Sci. USA 80:2462 (1983)
26. Konigsberg, W. H. and Henderson, L., Proc. Natl. Acad. Sci. USA 80:2467 (1983)
27. Amaya, Y., Kawamoto, S., Oda, T., Kuzumi, T., Saheki,T., Kimura, S. and Mori, M., Biochem. Int. in press
28. Kawamoto, S., Amaya, Y., Oda, T., Kuzumi, T., Saheki, T, Kimura, S. and Mori, M., Biochem. Biophys. Res. Commun. 136:955 (1986)

RED CELL ADENOSINE DEAMINASE OVERPRODUCTION ASSOCIATED WITH HEREDITARY HEMOLYTIC ANEMIA

Hisaichi Fujii
Hitoshi Kanno
Kenzaburo Tani
Shiro Miwa

Department of Internal Medicine
Institute of Medical Science
University of Tokyo
Tokyo, Japan

The mechanism of red cell adenosine deaminase (ADA) overproduction in hereditary hemolytic anemia due to increased red cell ADA activity was investigated. Red cell ADA from normal subjects and from a patient were purified using antibody affinity chromatography. We obtained 1.1 mg of the normal enzyme from 10 liters of normal blood samples and 1.4 mg of the patient's enzyme from 280 ml of the patient's blood. There were no differences in the molecular weight, specific activity, polyacrylamide-gel electrophoresis, Michaelis constant for substrate, thermal-stability, optimum pH, immunological reactivity, amino acid composition, and tryptic peptide mapping. These results strongly suggest that increased red cell ADA activity is caused by an accumulation of a structurally normal enzyme. The ADA activity in younger red cells was twice that of the older cells. Therefore, it is apparent that the increased number of the enzyme molecules is not subject to prolonged decay of ADA in the patient's red cell. Rate of ADA synthesis in erythroid colony cells cultured from the patient's bone marrow cells was 11-fold greater than that from the normal. The accumulation of ADA in the patient's red cell seems to be due to the increased synthesis in precursors of red cells. Furthermore, we examined the molecular basis of ADA overproduction in reticulocytes and spleen cells from the patient. Southern blot analysis of genomic DNA using ADA cDNA as a probe demonstrated no gene amplification or rearrangement. Dot blot analysis of reticulocyte RNA did not show increased amounts of ADA-specific mRNA. From these results, the defect might be mediated at the translational or post-translational rather than the transcriptional level.

I. INTRODUCTION

Adenosine deaminase (ADA) is an amino hydrolase which catalyzes the deamination of the purine riboside, adenosine, to produce inosine and ammonia. It is widely distributed in human tissues (1), and exists in different molecular and electrophoretic forms. Two forms have been characterized; the small form (molecular weight; 38,000) of the enzyme predominates in the spleen, stomach, and red cells, while the large form (molecular weight; 298,000) predominates in the kidney, liver, and skin fibroblasts. The small form can be converted to the large form by complexing with a 200,000-dalton protein known as the ADA binding protein. Clones encoding human ADA were isolated and characterized by three groups (2,3,4).

Low levels or the absence of ADA is associated with one form of severe combined immunodeficiency disease (4). On the other hand, increased ADA activity in red cells develops into hereditary hemolytic anemia. First family was reported by Valentine et al. in 1977 (5). The disease was trasmitted by an autosomal dominant trait. The enzyme activities were 45- to 70-fold of the normal, whereas the concentration of ATP in the red cell was about half that of reticulocyte-rich control. Subsequently, we discovered the second family of the same disorder (6), and a third case was reported in an Algerian child (7). Quite recently, we discovered the additional family in Japan. Only four such families have been discovered. Here we report the molecular basis of ADA overproduction associated with hemolytic anemia.

II. MATERIALS AND METHODS

Normal ADA was purified from 10 liters of 50 different ACD treated healthy donor blood samples obtained from individuals of the red cell ADA phenotype 1. The patient's enzyme was purified from 280 ml of heparinized blood collected from the patient. Preparation of antibody affinity Sepharose 4B and the method of purification were reported previously (8).

Normal and the patient's red cells were fractionated by centrifugation at 45,000 g for 1 hour. Younger and older cells were obtained from the top (about 10%) and the bottom (about 10%) of the tube, respectively.

Rate of ADA synthesis was examined by erythroid colony cell culture. Cell culture was carried out essentially according to the methyl cellulose technique described by Iscove et al. (9). Each 1 ml plate consisted of alpha medium containing $3X10^5$ mononuclear cells obtained from normal and the patient's bone marrow samples, 0.8% methyl cellulose, 30% fetal calf serum, 1% bovine serum albumin, $5X10^{-5}$M ß-mercaptoethanol, and 2 units of erythropoietin. The culture was grown at 37°C in 5% CO_2/95% air. On day 14 after the initiation of the culture, each plate was incubated with ^{3}H-leucine. After 24 hours of radioactive labelling, synthesized ADA was purified by the specific immunoprecipitation and SDS gel electrophoresis (10).

The ADA cDNA clone kindly provided by S.H.Orkin was used as radioactive probe. Plasmid DNA was extracted by using alkaline lysis procedure, and purified by CsCl/ ethidium bromide density gradient centrifugation. The PstI fragment, D1 reported previously (11), was purified by acrylamide gel electrophoresis, and nick-translated with (alpha-^{32}P)dCTP under conditions recommended by the manufacturer.
The isolation of high-molecular weight DNA from the spleen and Southern blotting were performed as described by Maniatis et al. (12). Poly(A)$^+$RNA from the reticulocyte was prepared according to the method reported by Temple et al. (13).
Dot blot analysis of reticulocyte RNA was performed as described by Thomas (14).

III. RESULTS AND DISCUSSION

The purification procedures are summarized in Table I. From these procedures, we obtained 1.1 mg of the normal enzyme and 1.4 mg of the patient's enzyme. The preparations were homogeneous on SDS-polyacrylamide gel electrophoresis. The molecular weight of the normal and the patient's enzymes were estimated to be 44,000 upon SDS-polyacrylamide gel electrophoresis. Polyacrylamide gel electrophoresis showed three bands using a Coomassie blue stain. Visualization of ADA on the slab gel electrophoresis also revealed three bands.

TABLE I. Purification of Adenosine Deaminase from Normal Red Cells

Step	Volume (ml)	Total Activity (U)	Total Protein (mg)	Specific Activity (U/mg)	Purification (fold)	Recovery (%)
Hemolysate	20,000.0	2,130	1,062,600	0.002	1	100
DEAE-Sephadex	4,870.0	2,326	8,279	0.281	140	109.2
60% $(NH_4)_2SO_4$	256.0	1,547	3,978	0.389	194	72.6
Sephadex G-75	3,050.0	1,300	2,236	0.581	291	61.0
60% $(NH_4)_2SO_4$	59.0	1,365	1,811	0.754	377	64.1
Antibody affinity	3.2	501	1.066	470	234,991	23.5

Purification of Adenosine Deaminase from Patient's Red Cells

Step	volume (ml)	Total Activity (U)	Total Protein (mg)	Specific Activity (U/mg)	Purification (fold)	Recovery (%)
Hemolysate	640.0	2,616	46,720	0.056	1	100
DEAE-Sephadex	370.0	2,476	348	7.11	127	94.6
60% $(NH_4)_2SO_4$	17.6	2,547	168	15.1	270	97.2
Sephadex G-75	120.0	1,667	113	14.7	263	63.7
60% $(NH_4)_2SO_4$	9.8	1,488	90	16.5	295	56.9
Antibody affinity	3.6	700	1.421	492	8,797	26.8

TABLE II. Enzymatic Properties of Normal and the Patient's Enzymes

	Normal	Patient
Specific activity (U/mg protein)	470	492
Km for adenosine (μM)	26.8	27.3
Ki for guanylurea (μM)	24.3	26.7
Utilization of 2-deoxyadenosine (% of adenosine)	79.6	76.3
Optimum pH	7.3	7.3

The electrophoretic mobility of the patient's enzyme was identical to that of the normal enzyme with each electrophoretic trial. The values of specific activity, Km for adenosine, Ki for guanylurea, utilization of 2-deoxyadenosine, and optimum pH are presented in Table II. The heat stability test and neutralization of enzyme activity using various dilutions of antiserum revealed no difference between the normal and the patient's enzymes. Amino acid composition and the tryptic peptide map of the patient's enzyme corresponded closely to those of the normal enzyme.

As mentioned above, we could observe no distinction between the normal and the patient's enzymes in enzymatic properties and also chemical characteristics. From enzymatic properties, we can conclude that the increased enzyme activity results from an increased amount of enzyme with normal specific activity and normal enzymatic properties. Moreover, by analogy with the chemical charactristics, in particular the pattern of peptide mapping, it is most conceivable that the increased ADA activity represents increased amounts of structurally normal enzyme.

Enzyme activities of younger and older cells are presented in Table III. With respect to pyruvate kinase (PK) and

TABLE III. Enzymatic Activities in Red Cells of Different Age Groups

	Normal		Patient	
	Younger	Older	Younger	Older
Reticulocyte count (%)	1.8	0.6	4.0	1.4
Adenosine deaminase (U/gHb)	0.35	0.31	33.0	17.0
Pyruvate kinase (U/gHb)	25.3	10.6	87.3	32.1
Hexokinase (U/gHb)	1.63	1.07	3.59	1.56

TABLE IV. Incorporation of Radioactive Leucine into Adenosine Deaminase during 24 Hours Incubation

	Incorporation ($cpm/10^5$ cells)
Control 1	1,008
Control 2	730
Reticulocyte-rich control*	2,236
Patient	9,644

* PK deficiency (reticulocyte count; 55.6%)

hexokinase, which are age-dependent enzymes, the younger cells of both normal control and the patient contained more than twice as much enzyme activity as the older cells. There was no significant difference in ADA activity between the normal younger and the normal older cells whereas ADA activity of the patient's younger cells was twice that of the patient's older cells. Incorporation of radioactive leucine into ADA during 24 hours incubation is shown in Table IV. Rate of ADA synthesis in erythroid colony cells cultured from the patient's bone marrow cells was 11.1-fold greater than that from the normal and 4.3-fold greater than that of the case of PK deficiency with marked reticulocytosis.

In general, increased enzyme activity may be due to an increased number of enzyme molecules with normal catalytic activity either due to increased rate of synthesis, or due to decreased breakdown or decreased post-synthetic change. Alternately there may be a normal number of enzyme molecules with increased catalytic activity. From findings reported above, the increased enzyme activity results from an increased amount of structurally normal enzyme with normal specific activity and normal enzymatic properties. The decay of ADA activity during the life of the red cell is very slow since there is no significant difference in activity between normal controls and controls with reticulocytosis. From the data of Table III, decreased ADA activity was present in the patient's older cells. Therefore, it is apparent that the increased number of the enzyme molecules is not subject to prolonged decay of ADA in his red cells. Isotope incorporation studies confirm that erythroid colony cells from the patient's bone marrow synthesized much more ADA than those of the normal. Thus, the accumulation of ADA seems to be due to the increased synthesis in precursors of red cells in spite of the increased degradation in peripheral blood.

By Southern blot hybridization, XbaI-digested DNA from

the normal control and the patient showed a 6.0-kb band which hybridized to the ADA cDNA probe. The patient did not show size-difference or an increased amount of ADA gene sequence. Therefore, we could not obtain the evidence of ADA gene amplification or rearrangement. Dot blot analysis of reticulocyte RNA also showed that the patient contained almost the same quantities of ADA-specific mRNA as compared to the normal control.

The results indicate that the defect might be mediated at the translational or post-translational rather than the transcriptional level. One possible mechanism is the abnormality of 5'-flanking region which affects translation efficiency. The sequence of these region of the patient's gene is necessary to disclose the precise mechanism of structurally normal enzyme overproduction.

ACKNOWLEDGMENTS

This study was supported in part by research grants from the Ministry of Education, Science and Culture and the Ministry of Health and Welfare, Japan.

REFERENCES

1. Edwards, Y. H., Hopkinson, D. A., and Harris, H., Ann. Hum. Genet., Lond. 35:207 (1971).
2. Orkin, S. H., Daddona, P. E., Shewach, D. S., Markham, A. F., Bruns, G. A., Goff, S. C., and Kelley, W. N., J. Biol. Chem. 258:12753 (1983).
3. Valerio, D., Duyvesteyn, M. G. C., Meera Kahn, P., Geurts van Kessel, A., de Woard, A., and van der Eb, A. J., Gene 25:231 (1983).
4. Giblett, E. R., Anderson, J. E., Cohen, F., Pollara, B., and Meuwissen, H. J., Lancet II:1067 (1972).
5. Valentine, W. N., Paglia, D. E., Tartaglia, A. P., and Gilsantz, F., Science 195:783 (1977).
6. Miwa, S., Fujii, H., Matsumoto, N., Nakatsuji, T., Oda, S., Asano, H., Asano, S., and Miura, Y., Am. J. Hematol. 5:107 (1978).
7. Pèrignon, J,-L., Hamet, M., Bue, H. A., Cartier, P. H., and Derycke, M., Clin. Chim. Acta 124:205 (1982).
8. Fujii, H., Miwa, S., and Suzuki, K., Hemoglobin 4:693 (1980).

9. Iscove, N. N., Sieber, F., and Winterhalter, K. H., J. Cell. Physiol. 83:309 (1974).
10. Fujii, H., Miwa, S., Tani, K., Fujinami, N., and Asano, H., Brit. J. Haematol. 51:427 (1982).
11. Orkin, S. H., Goff, S. C., Kelley, W. N., and Daddona, P. E., Mol. Cell. Biol. 5:762 (1985).
12. Maniatis, T., Fritsch, E., and Sambrook, J., in Molecular Cloning: A Laboratory Manual, Cold Spring Harbor Laboratory, New York, (1982).
13. Temple, G. F., Chang, J. C., and Kan, Y. W., Proc. Natl. Acad. Sci. USA 74:3047 (1972).
14. Thomas, P. S., Proc. Natl. Acad. Sci. USA 77:5201 (1980).

GENE TRANSFER AND EXPRESSION OF HUMAN PURINE NUCLEOSIDE PHOSPHORYLASE AND ADENOSINE DEAMINASE: POSSIBILITIES FOR THERAPEUTIC APPLICATION

R. Scott McIvor[1]
Sharon Pitts
David W. Martin, Jr.

Genentech, Inc.
South San Francisco, California

I. INTRODUCTION

Recent advances in gene transfer technology have made possible the insertion of new genetic material into somatic tissue of animals (Miller et al., 1984; Williams et al. 1984). One potential application of such procedures would be the treatment of genetic disease by gene supplementation therapy. Optimism that gene transfer techniques might be thus applied has been strengthened by the success of bone marrow transplantation as an approach to dealing with certain human genetic disorders (Parkman, 1986). This in itself constitutes a form of "gene therapy", in that the grafted material contains new genetic information which is different from that of the host.

Initial attempts at human gene supplementation therapy will most likely be in the treatment of monogenic disorders which result in simple metabolic deficiency (Anderson 1984). In this paper, the immunodeficiency diseases associated with the absence of purine nucleoside phosphorylase or adenosine deaminase are presented as candidate conditions for such treatment. Work is also described on the development and use of retroviral vectors designed for transfer and expression of these genes in mammalian cells and tissues.

[1]Present address: Institute of Human Genetics, University of Minnesota, Minneapolis, Minnesota.

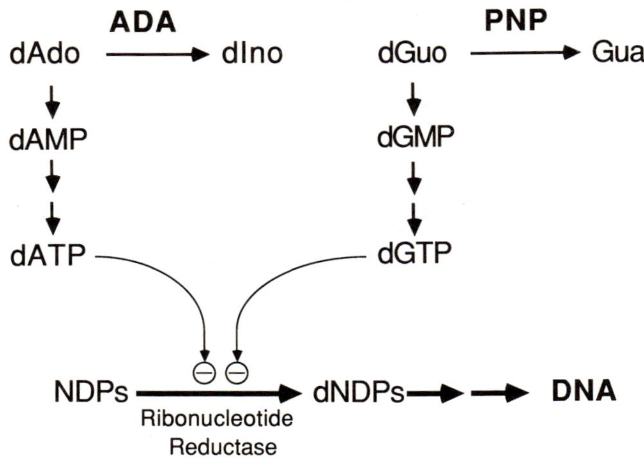

Figure 1. Metabolism and toxicity of deoxyadenosine (dAdo) and deoxyguanosine (dGuo). Also abbreviated are deoxyinosine (dIno), guanine (gua) and nucleoside diphosphates (NDP's).

II. ADA$^-$ and PNP$^-$ IMMUNODEFIENCY DISEASES AND GENE THERAPY

In humans, the absence of adenosine deaminase (ADA) activity is associated with severe combined immunodeficiency (Giblett et al., 1972). A T-cell immunodeficiency is observed in the absence of purine nucleoside phosphorylase (PNP) activity (Giblett et al., 1975). These rare conditions are inherited in an autosomal recessive manner and result in death of the affected individual early in childhood due to the inability to resist common infections (Martin and Gelfand, 1981).

The biochemical basis for these diseases appears to be associated with the accumulation of substrates (Martin and Gelfand, 1981). In the case of purine nucleoside phosphorylase there are four substrates (inosine, guanosine, deoxyinosine, and deoxyguanosine), all of which accumulate in the blood and urine of PNP-deficient individuals. In T-cells, deoxyguanosine can be phosphorylated to deoxyGMP and hence to deoxyGTP (Fig. 1). This increased intracellular level of deoxyGTP feedback inhibits ribonucleotide reductase leading to imbalanced deoxynucleoside triphosphate pools and DNA synthesis which is insufficient to support an immunoproliferative response (Ullman et al., 1979). It has been similarly proposed that immunotoxicity in ADA deficiency is caused by accumulated deoxyATP and feedback inhibition of ribonucleotide reductase. Although these hypotheses provide a metabolic explanation for the observed immunological defects, they do not explain why ADA deficiency affects both arms of the immune system while PNP deficiency affects only cellular immunity.

Current treatment of children suffering from ADA$^-$ or PNP$^-$ associated immunodeficiencies involves close surveillance for infections, enzyme therapy in the form of infused, normal irradiated erythrocytes, and bone marrow transplantation when an HLA-matched donor is available.

The availability of cloned ADA and PNP coding sequences (Valerio et al., 1984; Goddard et al., 1983) has led to the suggestion that these immunodeficiencies might respond to treatment by a gene therapy protocol in conjunction with autologous bone marrow transplantation (Anderson, 1984; Parkman, 1986). There are a number of reasons which support this idea. First, since these are diseases of the hematopoietic system, bone marrow would provide a natural target tissue which is relatively accessible and manipulable. Gene transfer into hematopoietic tissue of the mouse has been demonstrated using recombinant retroviral vectors (Miller et al., 1984; Williams et al., 1984). Secondly, for these diseases it would not be anticipated that the presence of the inactive, endogenous gene product would have any deleterious effect on the activity of a newly introduced and expressed gene or gene product, so displacement of the endogenous, dysfunctional gene should not be necessary. This defines the required therapy as gene supplementation. Third, although it may be necessary to obtain lymphoid cell- or organ-specific expression, it is possible that expression of active enzyme in any cell or lineage type would suffice in reversing the metabolic defect. The substrates and products of both PNP and ADA reactions are rapidly transported across cell membranes by facilitated diffusion (Plagemann and Wohlhueter, 1980). Therefore, the cell type in which expression is obtained might not make any difference, so long as sufficient enzyme activity is produced overall in the hematopoietic system to detoxify the deoxynucleoside substrates and prevent their accumulation. Finally, high levels of expression should not be necessary, since it has been observed that individuals that have only a few percent of the normal amount of ADA activity retain normal immune function (Martin and Gelfand, 1981). In summary, modification of current bone marrow transplantation procedures to include gene supplementation in the treatment of these diseases would be straightforward, and the success of such a procedure might not depend on highly regulated or tissue specific gene expression.

III. RETROVIRUSES TRANSDUCING ADA AND PNP GENE SEQUENCES.

Much of the current excitement about prospects for human gene therapy has been fueled by recent advances in the use of recombinant retroviruses as vectors for high efficiency gene transfer (Bernstein et al., 1985; Shimotohno and Temin, 1981; Williams et al., 1984). As gene transfer vehicles, retroviruses are attractive for a number of reasons, including stable integration of the proviral sequence as a part of the retroviral replicative cycle and the predictability of integrated proviral structures (Bernstein et al., 1985; Varmus 1982). Cis- and trans-acting requirements for the generation of recombinant replication-defective retroviruses have been fairly well established (Bernstein et al., 1985; Mann et al., 1983) and the use of amphotropic packaging systems provides recombinant viruses capable of infecting human cells (Cone and Mulligan, 1984; Miller et al., 1985). Demonstration of retroviral-mediated gene transfer into hematopoietic stem cells of the mouse (Dick et al., 1985; Eglitis et al., 1985; Keller et al., 1985; Lemischka et al., 1986; Miller et al., 1984; Williams et al., 1984; Williams et al., 1986) is particularly pertinent to the possible use of these vectors in treating genetic diseases affecting the hematopoietic system.

An initial set of ADA- and PNP-retroviruses were constructed which contained either coding sequence just downstream from the mouse metallothionein promoter (McIvor et al., manuscript submitted). These viruses also contained a sequence encoding human hypoxanthine phosphoribosyltransferase as a mammalian selectable marker; this gene was used for establishing virus-producer cell lines and for determining virus titer. In these initial studies, it was found that some of these viruses underwent a deletion which was the result of splicing (McIvor and Martin, manuscript in preparation). ADA- and PNP-coding sequences remained intact, however, and active human enzyme was detected in mouse cells infected with these viruses. These recombinant viruses were also used to introduce ADA and PNP gene sequences into hematopoietic stem cells of the mouse (section IV).

Employing the HPRT gene as a selectable marker for establishing virus-producer cell lines required the use of an HPRT-deficient 3T3 cell line as a target cell population, with subsequent co-transfection with a packaging sequence (McIvor et al., mansucript submitted). A more convenient method for establishing virus producer cell lines is to use a dominant selectable marker (Miller et al., 1986). Four ADA-viruses are depicted in Fig. 2 which were constructed to

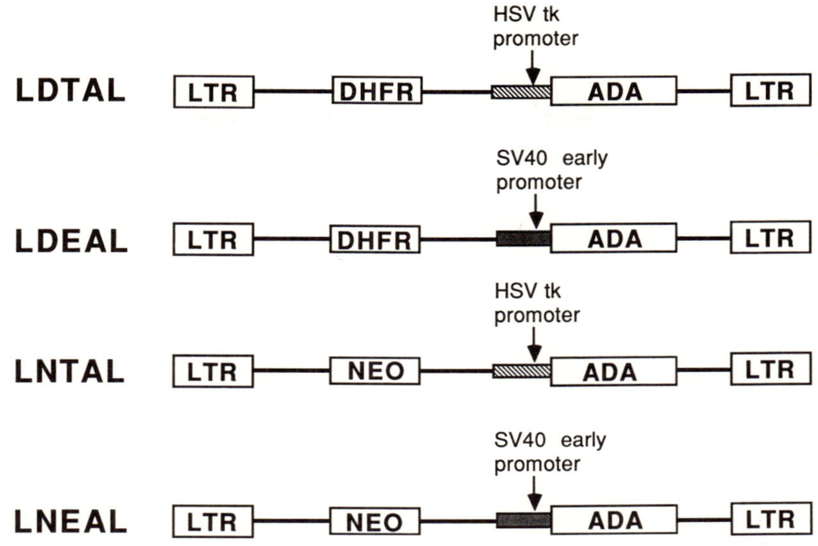

Figure 2. ADA-retrovirus constructions inserted into pBR322. 5' and 3' LTR's (long terminal repeats) were from Moloney sarcoma virus and Moloney murine leukemia virus (MoMLV), respectively. Other viral sequence is indicated by the solid line. The DHFR sequence (Simonsen and Levinson, 1983) was a 728 bp PstI-SstII fragment from pMPMD (McIvor et al., 1985). The NEO sequence was a 986 bp HindIII-SmaI fragment from pSVEneoBal6 (Seeburg et al., 1984). The ADA coding region was a 1358 bp BamHI-SstII sequence derived from pMAMD (Valerio et al., 1984). The SV40 early promoter was inserted as a 342 bp PvuII-HindIII fragment from pEPD (McIvor et al., 1985) between a MoMLV EcoRI site on the 5' side and a HindIII site on the 3' side, just upstream from the ADA gene. The HSV TK promoter (McKnight, 1980) was either a 251 bp PvuII-BglII fragment (LDT$_s$AL; "s" for short) or a 434 bp AccI-BglII fragment (LNTAL and LDT$_l$AL; "l" for long) inserted between the MoMLV EcoRI site and a SalI site just upstream from the ADA gene.

contain either the bacterial neomycin-resistance gene (conferring resistance to the antibiotic G-418, a neomycin analog, in mammalian cells) or a sequence encoding a methotrexate-resistant dihydrofolate reductase within the context of a retroviral genome. The herpes simplex virus 1 thymidine kinase (TK) promoter or the SV40 early promoter (Benoist and Chambon, 1981) were also included just upstream from the ADA gene. Plasmid DNA was transfected (Gorman, 1985) into the amphotropic packaging line PA12 (Miller, et al., 1985) to generate transient, low titer recombinant virus (collected 2 days post-transfection). This was used to infect psi-2 ecotropic packaging cells (Mann et al., 1983) during an overnight incubation. The infected psi-2 cells were then plated into selective medium (containing either 0.4 mg/ml G-418 or 0.25 uM methotrexate and 7% dialyzed fetal calf serum). Several clones of each virus type were picked after 14 days, expanded, and those cultures identified which produced the highest titer of virus in the culture medium during an overnight exposure.

Table 1 summarizes the virus titers obtained for the 4 ADA-virus constructs (Fig. 2) after shuttling into psi-2. The viruses containing the NEO gene (LNTAL, LNEAL) provided higher titers than the DHFR-viruses (LDEAL, LDT_sAL, LDT_1AL). This might have been due in part to inherent differences between the two selective systems. However, we observed that under non-selective conditions the NEO-viruses provided a higher integration frequency than the DHFR viruses in spleen colonies (see below), indicating a genuine difference in the number of infectious particles being produced. Slightly higher titers were observed for viruses containing the SV40 early promoter than for those containing the TK promoter (Table 1).

To determine whether the structure of these viruses was maintained during the replicative cycle, DNA was extracted from the virus-producer lines and subjected to Southern analysis (Southern, 1975). SstI digests were electrophoretically fractionated, blotted onto nitrocellulose, and probed for human ADA sequences (Fig. 3). SstI cleaves in

Table 1. ADA-virus titers.

Virus	No. of cell lines tested	Titer (cfu/ml)[1]
LDEAL	2	5.9×10^5
LDT_1AL	6	5.5×10^5
LDT_sAL	7	2.3×10^5
LNEAL	6	2.6×10^6
LNTAL	5	1.8×10^6

[1] Number of particles per ml of overnight culture medium which genetically transform NIH 3T3 tk^- cells to G-418 or methotrexate (0.1 uM) resistance.

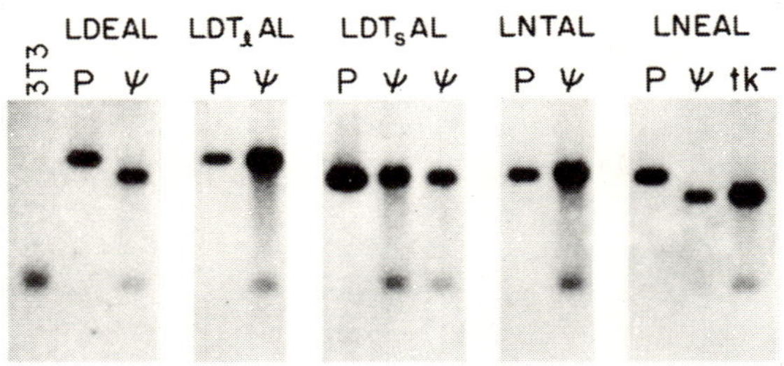

Figure 3. Southern analysis of ADA-virus-infected cells. Plasmid DNA (P) or DNA extracted from psi-2 producer cell lines (Ψ) for each of the ADA-viruses (see Fig. 1) was digested to completion with SstI. DNA extracted from uninfected 3T3 cells was included as a negative control. DNA from a heterogeneous population of LNEAL-infected and G-418-selected 3T3 tk$^-$ cells (tk$^-$) was also included. Digests were electrophoretically fractionated, blotted onto nitrocellulose, and probed for human ADA sequences using a 1194 bp BamHI fragment derived from pMAMD (Valerio et al., 1984) labeled by extension of random primers (McIvor et al., 1985).

the LTR regions of these viruses (see Fig. 2), thus generating genome-length fragments. All of the ADA-viruses which included the TK promoter (LDT$_1$AL, LDT$_S$AL, and LNTAL) generated ADA-hybridizing SstI fragments which co-migrated with those of the original plasmid constructions. There were thus no major changes in the structure of these sequences through the one replicative cycle required for shuttle packaging (see above). However, the ADA-viruses which included the SV40 early promoter (LDEAL, LNEAL) generated ADA-hybridizing SstI fragments which were shorter than those obtained from the original plasmid constructs. A portion of the recombinant retroviral sequence appears to have been deleted from these viruses. These results are quite similar to those observed with viruses constructed to contain an 1834 bp mouse metallothionein promoter sequence just upstream from either the ADA or the PNP gene (McIvor et al., manuscript submitted). One of these proviral sequences (LHM$_1$PL) has been molecularly cloned, and sequencing across the deletion junctions has indicated the presence of cryptic splice donor and acceptor sites contained in the original recombinant viral construct which were responsible for the observed deletion (McIvor and Martin, manuscript in preparation). We are currently undertaking experiments to determine whether splicing was responsible for the deletion in LNEAL as well.

Interestingly, the viruses containing the TK promoter expressed a higher level of ADA activity than those containing the SV40 early promoter when crude cell extracts were assayed by isozyme analysis on isoelectric focusing gels (data not shown).

IV. PNP AND ADA GENE TRANSFER INTO MURINE HEMATOPOIETIC STEM CELLS

In evaluating the potential use of a vector for gene transfer and expression in human hematopoietic tissue, it would be useful to demonstrate its effectiveness in an animal model. Application of retroviral-mediated gene transfer to the mouse spleen colony forming assay (Williams et al., 1984) has provided a convenient method of obtaining genetically transformed hematopoietic tissue which is derived from a relatively primitive stem cell (Till and McCullough, 1964, 1980). We administered velban (3 mg/kg i.p.) to donor C3H or DBA mice 2 days prior to bone marrow harvest to stimulate progenitor cell proliferation (Smith et al., 1977) and thereby increase the susceptibility of these cells to retroviral integration (Varmus and Swanson, 1983; discussed in Lemischka et al., 1986 and in McIvor et al., manuscript submitted). Bone marrow was flushed from the hind limbs of donor animals and a relatively small number of the marrow cells were then co-cultured overnight with virus-producer cells. The supernatant cells were collected and injected (2×10^5 cells) into lethally irradiated (700 rads X-ray) syngeneic recipient mice. Spleens were removed 10-12 days later. Nodules (spleen colonies) formed on spleens from the clonal outgrowth of a single hematopoietic stem cell in the original bone marrow population (Till and McCullough 1964) were excised and a portion extracted for DNA. Crude cell extracts were also prepared for isozyme analysis (see below).

Initial animal experiments were undertaken using amphotropic retroviruses originally constructed to contain the mouse metallothionein promoter just upstream from either the PNP or the ADA gene (see above). Freshly explanted bone marrow cells were incubated overnight with virus-producer cells and then injected intravenously into lethally irradiated recipient mice. Southern analysis of DNA extracted from subsequently formed spleen colonies demonstrated the presence of ADA- or PNP-proviruses in 36-55% of the colonies analyzed (McIvor et al., manuscript submitted). However, crude cell extracts of spleen colony tissue positively identified for the presence of ADA-proviral sequences did not contain detectable human ADA activity when assayed on isoelectic focusing gels. Since the metallothionein promoter element originally positioned just upstream from the ADA gene had been deleted, ADA expression in virus-infected fibroblasts most likely depended on the LTR for initiation of transcription, and it appeared that this organization was not functional in spleen colony tissue.

We were interested in determining the effect of an intact internal promoter in the proviral sequence integrated into hematopoietic cells _in vivo_. The herpesvirus thymidine kinase promoter was of particular interest because this element has been shown to efficiently inititate transcription after gene transfer into mouse embryonal carcinoma stem cells (Wagner et al., 1985). The ADA-viruses containing the TK promoter described above (LDTAL and LNTAL, Fig. 2) were structurally stable through the retroviral replicative cycle (Fig. 3) and thus provided a good vector for testing the role of an internal promoter on ADA expression in spleen colony tissue.

All four of the ADA-viruses shown in Fig. 2 were introduced into mouse spleen colony tissue using the procedure described above. The percentage of spleen colonies in which ADA-virus sequences were detected (by Southern analysis; data not shown) varied directly with the titer of virus obtained from the specific producer line employed (Table 1) (from 10% for LDT_1AL to 75% for LNEAL). However, human ADA activity was not detected in any of these spleen colony samples when screened by isozyme analysis.

In addition to facilitating the establishment of virus-producer cell lines, the DHFR and NEO genes can be used as selectable markers in stem cells. Retroviral-mediated transfer of either gene into hematopoietic progenitors can allow _in vitro_ colony formation in the presence of the appropriate selective agent (G-418 or methotrexate) (Hock and Miller, 1986; Joyner et al., 1983). Dick et al. (1985) demonstrated the use of the NEO gene in a preselective procedure to eliminate CFU-s (colony forming units - spleen; those stem cells which give rise to spleen colonies) which have failed to integrate and express a NEO-virus prior to transplantation.

The NEO-selection procedure of Dick et al. was applied using LNTAL as a gene transfer vector. Bone marrow cells co-cultivated overnight with LNTAL-producer cells (procedure described above) were collected and incubated for 2 days in medium containing 2 mg/ml G-418. Injection of the cells into irradiated recipient mice resulted in the formation of a very small number of spleen colonies (0.5 per 10^5 cells in the original bone marrow population, compared with a transplantation frequency of 12 colonies per 10^5 cells under non-selective conditions) indicating that the selective procedure had been effective. Southern analysis of DNA

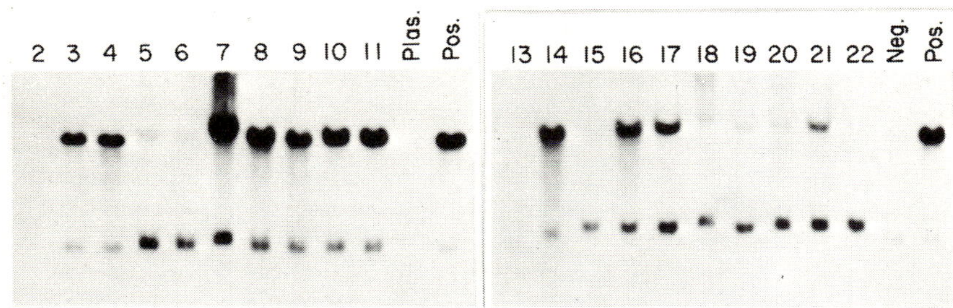

Figure 4. Southern analysis of DNA extracted from mouse spleen colonies. Samples 3, 4 and 7-16 were from G-418 selected material while samples 5, 6 and 17-22 were from unselected material (see text). DNA extracted from the LNTAL-producer cell line (Pos.) and from normal, uninfected spleen colony tissue (Neg.) were included as controls. See Fig. 3 for blotting and probing procedure. Sample 2 was not scored in the text since it did not appear to contain any DNA (indicated by lower molecular weight endogenous ADA-hybridizing band).

extracted from preselected colonies indicated the presence of ADA-virus sequences in 9 out of 11 samples (Fig. 4), 82%. Those positive samples (3,4,7-11,14 and 16) contained a much stronger ADA-hybridizing signal than samples derived from infected but non-selected marrow (samples 5,6,17-22), suggesting a higher viral integration frequency in selected material. However, none of the samples, selected or non-selected, contained human ADA activity which was distinguishable from minor isozymes of mouse ADA by isoelectric focusing (Fig. 5). Thus, the TK promoter was ineffective in providing expression of detectable human ADA activity within the genetic context of this murine retrovirus system even when selective conditions were employed.

V. CONCLUSIONS

The recombinant ADA- and PNP-retroviruses described above were designed to provide high-efficiency vectors for the transfer and expression of these genes in mammalian cells and, ultimately, in mammalian (especially human) hematopoietic tissue. Although several viruses effectively integrated either PNP or ADA gene sequences into CFU-s of the mouse (e.g. Fig. 4), none expressed detectable human enzyme activity in spleen colony tissue (e.g. Fig. 5). These

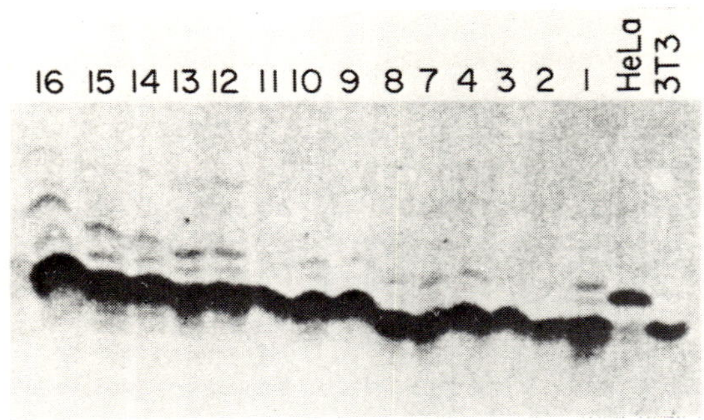

Figure 5. ADA isozyme analysis of crude cell extracts from spleen colony tissue (samples summarized in Fig. 4). Extracts were prepared by three cycles of freeze-thawing and subjected to isoelectric focusing between pH 4 and pH 5. The gel was then stained *in situ* for ADA activity (see Valerio et al., 1984 for procedure). The gel is oriented with the cathode at the top. Human ADA (HeLa) focuses at pI 4.6, while mouse ADA (3T3) focuses at pI 4.4. Also observed are several minor mouse ADA isozymes contained in spleen colony tissue.

viruses either contained an internal HSV thymidine kinase promoter or depended on the MoMLV LTR for initiation of transcription. Since the same viruses expressed substantial human ADA activity in mouse fibroblasts, there appears to be some constraint on ADA expression in spleen colony tissue which does not function in the cultured cells. This has also been observed in spleen colonies harboring viruses which contain the SV40 early promoter (Williams et al., 1986) or the human ADA promoter (Valerio et al., Proceedings of the 2nd Meeting of the European Group on Immune Deficiencies, in press) just upstream from the ADA coding sequence. The nature of this expressional constraint might be related to the stage of differentiation of the target cell, since ADA expression has been demonstrated after retroviral-mediated gene transfer in mouse hematopoietic colonies *in vitro* (Belmont *et al*., 1986) and in lymphoblast cultures (Friedman, 1985; Williams et al., 1986).

Recombinant retroviruses can be used to transduce gene sequences into primitive stem cells capable of extensive proliferation and tissue regeneration in the mouse (Dick et al., 1985; Keller et al., 1985, Lemischka et al., 1985; Williams et al., 1984). Such long-term survival of target cells and their progeny would be necessary for a gene therapy protocol to effect a lasting cure. Identification of molecular issues pertinent to expression of genes newly introduced into hematopoietic stem cells will assist the application of gene transfer techniques to the treatment of genetic disease. The ADA^- and PNP^- immunodeficiencies are good candidate diseases for such application due to the characteristics of their genetic and biochemical pathology (section II). Proceeding to gene therapy will require a risk-benefit analysis that takes into account the proven efficacy of a specific vector (in an animal model) as well as safety concerns associated with the gene transfer process. These factors must be weighed carefully against the patients' chances for survival given currently available treatment, since untreated these diseases are fatal (Martin and Gelfand, 1981).

ACKNOWLEDGMENTS

We are grateful to Dr. Inder M. Verma for valuable discussion, to Mike Cagle for technical assistance, to Wayne Anstine for preparing the manuscript, and to Dr. Ingrid Caras for reviewing the manuscript.

REFERENCES

Anderson, W.F. (1984). Science 226:401.

Belmont, J.W., Henkel-Tigges, J., Chang, S.M.W., Wager-Smith, K., Kellems, R.E., Dick, J.E., Magli, C.M., Phillips, R.A., Bernstein, A., and Caskey, C.T. (1986). Nature 322:385.

Benoist, C., and Chambon, P. (1981). Nature 290:304.

Bernstein, A., Berger, S., Huszar, D., and Dick, J. (1985). In "Genetic Engineering: Principles and Methods". (K. Setlow and A. Hollaender, eds.), vol. 7, p. 235. Plenum Publishing Corporation.

Cone, R.D. and Mulligan, R.C. (1984). Proc. Natl. Acad. Sci. U.S.A. 81:6349.

Dick, J.E., Magli, M.C., Huszar, D., Phillips, R.A., and Bernstein, A. (1985). Cell 42:71.

Eglitis, M.A., Kantoff, P., Gilboa, E., and Anderson, W.F. (1985). Science 230:1395.

Friedman, R.L. (1985). Proc. Natl. Acad. Sci. U.S.A. 82:703.

Giblett, E.R., Ammann, A.J., Wara, D.W., Sandman, R. and Diamond, L.K. (1975). Lancet I:1010.

Giblett, E.R., Anderson, J.E., Cohen, F., Pollara, B., and Meuwissen, H.J. (1972). Lancet II:1067.

Goddard, J.M., Caput, D., Williams, S.R., and Martin, D.W., Jr. (1983). Proc. Natl. Acad. Sci. USA 80:4281.
Gorman, C. (1985). In "DNA Cloning" vol. II, (D. Glover, ed.), p. 143, IRL Press, Oxford, Washington.
Hock, R.A., and Miller, A.D. (1986). Nature 320:275.
Joyner, A., Keller, G., Phillips, R.A., and Bernstein, A. (1983). Nature 305:556.
Keller, G., Paige, C., Gilboa, E., and Wagner, E.F. (1985). Nature 318:149.
Lemischka, I.R., Raulet, D.H., and Mulligan, R.C. (1986). Cell 45:917.
Mann, R., Mulligan, R.C., and Baltimore, D. (1983) Cell 33:153.
Martin, D.W., Jr., and Gelfand, E.W. (1981). Ann. Rev. Biochem. 50:845.
McKnight, S.L. (1980) Nucl. Acids Res. 8:5949.
McIvor, R.S., Goddard, J.M., Simonsen, C.C., and Martin, D.W., Jr. (1985). Mol. Cell. Biol. 5:1349.
Miller, A.D., Eckner, R.J., Jolly, D.J., Friedmann, T., and Verma, I.M. (1984). Science 225:630.
Miller, A.D., Law, M.-F., and Verma, I.M. (1985). Mol. Cell. Biol. 5:431.
Miller, A.D., Trauber, D.R., and Buttimore, C. (1986). Som. Cell Mol. Genet. 12:175.
Parkman, R. (1986). Science 232:1373.
Plagemann, P.G.W., and Wohlhueter, R.M. (1980) Curr. Topics Membr. Transp. 14:225.
Seeburg, P.H., Colby, W.W., Capon, D.J., Goeddel, D.V., and Levinson, A.D. (1984) Nature 312:71.
Shimotohno, K., and Temin, H.M. (1981). Cell 26:67.
Simonsen, C.C., and Levinson, A.D. (1983). Proc. Natl. Acad. Sci. USA 80:2495.
Smith, W.W., Wilson, S.M., and Fred, S.S. (1968). J. Natl. Cancer Inst. 40:847.
Southern, E. (1975). J. Mol. Biol. 98:503.
Stewart, T.A., Pattengale, P.K. and Leder, P. (1984). Cell 38:627.
Till, J.E., and McCullough, E.A. (1961). Radiation Res. 14:213.
Till, J.E., and McCullough, E.A. (1980). Biochim. Biophys. Acta 605:431.
Ullman, B., Gudas, L.J., Clift, S.M., and Martin, D.W., Jr. (1979). Proc. Natl. Acad. Sci. USA 76:1074.
Valerio, D., McIvor, R.S., Williams, S.R., Duyvesteyn, M.G.C., Van Ormondt, H., Van der Eb, A.J., and Martin, D.W., Jr. (1984). Gene 31:137.
Varmus, H. (1982). Science 216:812.

Varmus, H., and Swanson, R. (1982). In "RNA Tumor Viruses" (R. Weiss, N. Teich, H. Varmus, and J. Coffin eds.), p. 369. Cold Spring Harbor Laboratory, Cold Spring Harbor, NY.

Wagner, E.F., Varek, M. and Vennstrom, B. (1985). EMBO J. 4:663.

Williams, D.A., Lemischka, I.R., Nathan, D.G., and Mulligan, R.C. (1984). Nature 310:476.

Williams, D.A., Orkin, S.H. and Mulligan, R.C. (1986). Proc. Natl. Acad. Sci. USA 83:2566.

Index

D

E

U

V

X